Asa Gray

Botany for young people and common schools

How plants grow

Asa Gray

Botany for young people and common schools
How plants grow

ISBN/EAN: 9783337271657

Printed in Europe, USA, Canada, Australia, Japan

Cover: Foto ©berggeist007 / pixelio.de

More available books at **www.hansebooks.com**

Botany for Young People and Common Schools.

HOW PLANTS GROW,

A SIMPLE INTRODUCTION TO STRUCTURAL BOTANY.

WITH

A POPULAR FLORA,

OR AN ARRANGEMENT AND DESCRIPTION OF COMMON PLANTS, BOTH WILD AND CULTIVATED.

ILLUSTRATED BY 500 WOOD ENGRAVINGS.

By ASA GRAY, M.D.,

FISHER PROFESSOR OF NATURAL HISTORY IN HARVARD UNIVERSITY.

NEW YORK:

IVISON, BLAKEMAN, TAYLOR, & CO.,

138 & 140 GRAND STREET.

1881.

BOTANY FOR YOUNG PEOPLE.

Part First.

HOW PLANTS GROW.

> CONSIDER THE LILIES OF THE FIELD, HOW THEY GROW : THEY TOIL NOT, NEITHER DO THEY SPIN : AND YET I SAY UNTO YOU, THAT EVEN SOLOMON IN ALL HIS GLORY WAS NOT ARRAYED LIKE ONE OF THESE. — Matthew vi. 28, 29.

OUR LORD'S direct object in this lesson of the Lilies was to convince the people of God's care for them. Now, this clothing of the earth with plants and flowers — at once so beautiful and so useful, so essential to all animal life — is one of the very ways in which HE takes care of his creatures. And when Christ himself directs us to consider with attention the plants around us, — to notice how they grow, — how varied, how numerous, and how elegant they are, and with what exquisite skill they are fashioned and adorned, — we shall surely find it profitable and pleasant to learn the lessons which they teach.

Now this considering of plants inquiringly and intelligently is the study of BOTANY. It is an easy study, when pursued in the right way and with diligent attention. There is no difficulty in understanding how plants grow, and are nourished by the ground, the rain, and the air; nor in learning what their parts are, and how they are adapted to each other and to the way the plant lives. And any young person who will take some pains about it may learn to distinguish all our common plants into their kinds, and find out their names.

Interesting as this study is to all, it must be particularly so to Young People. It appeals to their natural curiosity, to their lively desire of knowing about things: it calls out and directs (i. e. educates) their powers of observation, and is adapted to sharpen and exercise, in a very pleasant way, the faculty of discrimination. To learn *how to observe* and *how to distinguish things* correctly, is the greater part of education, and is that in which people otherwise well educated are apt to be surprisingly deficient. Natural objects, everywhere present and endless in variety, afford the best field for practice; and the study when young, first of Botany, and afterwards of the other NATURAL SCIENCES, as they are called, is the best training that can be in these respects. This study ought to begin even before the study of language. For to distinguish *things* scientifically (that is, carefully and accurately) is simpler than to distinguish *ideas*. And in NATURAL HISTORY * the learner is gradually led from the observation of things, up to the study of ideas or the relations of things.

This book is intended to teach Young People how to begin to read, with pleasure and advantage, one large and easy chapter in the open Book of Nature; namely, that in which the wisdom and goodness of the Creator are plainly written in the VEGETABLE KINGDOM.*

---

* *Natural History* is the study of the productions of the earth in their natural state, whether minerals, plants, or animals. These productions make up what are called the *Three Kingdoms of Nature*, viz.: —

1. *The Mineral Kingdom*, which consists of the Minerals (earths, metals, crystals, &c.), bodies not endowed with life.

2. *The Vegetable Kingdom*, which comprehends Vegetables or Plants.

3. *The Animal Kingdom*, which comprehends all Animals.

The natural history of the mineral kingdom is named MINERALOGY.

The natural history of the vegetable kingdom is BOTANY, — the subject of this book.

The natural history of the animal kingdom is named ZOÖLOGY.

In the FIRST PART of this book we proceed to consider, under four principal heads or chapters, —

The SECOND PART of the book consists of a Popular Flora for Beginners, viz. a Classification and Description (according to the Natural System) of the Common Plants of the country, both Wild and Cultivated.

Then follows a Dictionary of the peculiar terms which we have occasion to use in describing plants, or their parts, combined with a full Index to Part I. Every science, and every art or occupation, has terms or technical words of its own, and must have them. Without them, all would be confusion and guess-work. In Botany the number of technical words which a young student need to know is by no means great, and a little diligent study and practice will make them familiar.

The first and most important thing for the student is, to know well the general plan of a plant and the way it grows ; the parts plants consist of; the uses of the several parts ; their general forms, and the names which are used to distinguish them. This is all very interesting and very useful in itself; and it is indispensable for studying plants with any satisfaction or advantage to find out their names, their properties, and the family they belong to ; i. e. to ascertain the kinds of plants.

Let the learners, or the class under their teacher, therefore, in the first place go carefully once through the First Part of the book, or at least through the first two chapters, verifying the examples and illustrations given, as far as possible, with their own eyes, and searching for other examples in the plants and flowers around them. Then they may begin to study plants by the Flora, or Second Part of the book, according to the directions given in the last section of Chapter IV. Whenever they meet with a word which they do not remember or clearly understand, they will look it out in the Index, and refer back to the place in the first part of the book where it is used and fully explained. Remember that every one has to creep before he can walk, and to walk before he can run. Only begin at the beginning; take pains to understand things as you go on, and cultivate the habits of accuracy and nice discrimination which this study is eminently adapted to inspire. Then each step will render the next one easy; you will soon make more rapid progress; will be able to ascertain with facility the names and the structure of almost all common plants; and will gradually recognize the various and interesting relationships which bind the members of the vegetable creation together in natural families, — showing them to be parts of one system; varied expressions, as it were, of the thoughts of their Divine Author; planned in reference to one another; and evidently intended to enlarge and enlighten our minds, as well as to gratify our senses, and nourish, clothe, warm, and shelter our bodies. So the study of Botany — the most fascinating branch of Natural History, especially for the young — becomes more and more interesting the more we learn of it, and affords a constant and unalloyed intellectual gratification.

When young students have thoroughly mastered this little book, they will be well prepared to continue the study in the *Lessons in Botany and Vegetable Physiology*, and in the *Manual of the Botany of the Northern United States*, by the same author.

The illustrations are referred to throughout by numbers, with "Fig." prefixed. The numbers occasionally introduced, within parenthesis-marks, and without any prefix, (as on p. 25, line 1, and p. 36, line 9,) are references to former paragraphs, where the subject, or the word used, has already been explained.

⁎ The illustrations on the first page represent: — Fig. 1. Our commonest wild species of true Lily, viz. the Canada Lily. Fig. 2. The Chalcedonian Lily, a native of Palestine, with scarlet flowers, supposed to be "The Lily of the Field" to which our Saviour referred in the Sermon on the Mount. Fig. 3. Lilies of the Valley, not true Lilies, but belonging to the Lily Family.

CHAPTER I.

SECTION I. — The Parts of a Plant.

1. PLANTS are chiefly made up of three parts, namely, of *Root*, *Stem*, and *Leaves*. These are called the plant's *Organs*, that is, its instruments. And as these parts are all that any plant needs for its growth, or vegetation, they are called the ORGANS OF VEGETATION.

2. Plants also produce *Flowers*, from which comes the *Fruit*, and from this, the *Seed*. These take no part in nourishing the plant. Their use is to enable it to give rise to new individuals, which increase the numbers of that kind of plant, to take the place of the parent in due time, and keep up the stock ; that is, to reproduce and perpetuate the species. So the Flower with its parts, the Fruit, and the Seed, are called the plant's ORGANS OF REPRODUCTION.

3. The different sorts of Lilies represented on the first page, and the common Morning-Glory on this page, show all the parts.

4. The Root (Fig. 4, *r*) is the part which grows downwards into the ground, and takes in nourishment for the plant from the soil. It commonly branches again and again as it grows : its smaller branches or fibres are named *Rootlets*. Real roots never bear leaves, nor anything besides root-branches or rootlets.

5. The Stem (Fig. 4, *s*) is the part which grows upwards, and bears the leaves and blossoms. At certain fixed places the stem bears a leaf or a pair of leaves.

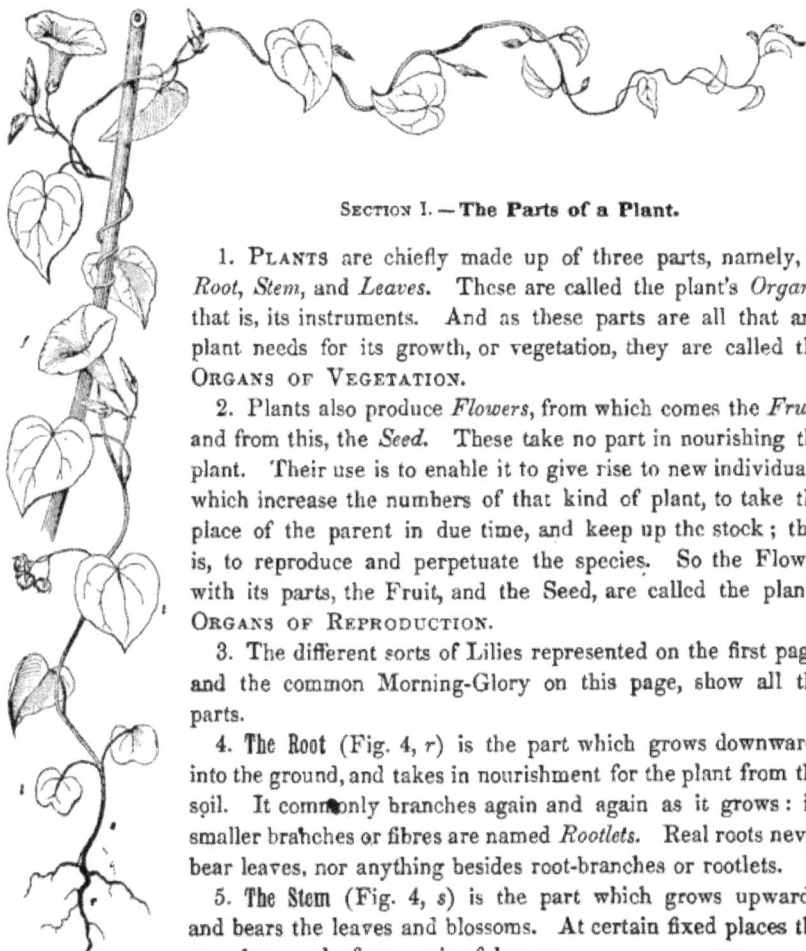

4. Morning-Glory.

6. **Leaves** (Fig. 4, *l, l*) are generally flat and thin, green bodies, turning one face upwards to the sky, and the other downwards towards the ground. They make the *Foliage*.

7. **The Plant in Vegetation.** We see that a plant has a body or trunk (in scientific language, an *axis*), consisting of two parts, — an upper and a lower. The lower is the Root: this fixes the plant to the soil. The upper is the Stem: this rises out of the ground, and bears leaves, which are hung out on the stem in the light and air. The root takes in a part of the plant's food from the soil: this the stem carries to the leaves. The leaves take in another part of the plant's food from the air. And in them what the roots absorb from the ground, and what they themselves absorb from the air, are exposed to the sunshine and *digested*; that is, changed into something proper to nourish the plant. For there is no nourishment in earth, air, and water as they are; but vegetables have the power of making these into nourishment. And out of this nourishment it prepares, the plant makes more growth. That is, it extends the roots farther into the soil, and sends out more branches from them, increasing its foothold and its surface for absorbing; while, above, it lengthens the stem and adds leaf after leaf, or shoots forth branches on which still more leaves are spread out in the light and air.

8. So the whole herb, or shrub, or tree, is built up. A tiny herb just sprouted from the seed and the largest tree of the forest alike consist of root, stem, and leaves, and nothing else. Only the tree has larger and more branching stems and roots, and leaves by thousands.

9. **The Plant in Reproduction.** After having attended in this way to its nourishment and growth for a certain time, the plant sets about reproducing itself by seed. And for this purpose it *blossoms*. Many plants begin to blossom within a few weeks after springing from the seed. All our *annuals*, of which the Garden Morning-Glory (Fig. 4) is one, blossom in the course of the summer. *Biennials*, such as the Carrot, Parsnip, Mullein, and the common Thistle, do not flower before the second summer; and shrubs and trees, and some herbs, do not begin until they are several years old.

10. The object of the Flower is to form the Fruit. The essential part of the fruit is the Seed. And the essential part of a seed is the *Germ* or *Embryo* it contains. The Germ or Embryo is a little plantlet in the seed, ready to grow into a new plant when the seed is sown. Let us notice these organs one after the other, beginning with

11. The Flower. Flowers are most interesting to the botanist; who not only admires them for their beauty, the exquisite arrangement and forms of their parts, and the wonderful variety they exhibit, but also sees in the blossoms much of the nature or character of each plant, and finds in them the best marks for distinguishing the sorts of plants and the family they belong to. So let the student learn at once

12. What the Parts of a Flower are. A flower, with all the parts present, consists of *Calyx, Corolla, Stamens,* and *Pistils.* One from the Morning-Glory (Fig. 4, *f*) will serve for an example. Here is one taken off, and shown of about the natural size, the corolla, Fig. 5, separated from the calyx, Fig. 6. The calyx and the corolla are the *Floral Envelopes,* or the leaves of the flower. They cover in the bud, and protect the stamens and pistils, which are the *Essential Organs* of the flower, because both of these are necessary to forming the seed.

13. The Calyx — a Latin name for "flower-cup" — is the cup or outer covering of the blossom (Fig. 6). It is apt to be green and leaf-like.

14. The Corolla is the inner cup, or inner set of leaves, of the flower. It is very seldom green, as the calyx commonly is, but is "colored," i. e. of some other color than green, and of a delicate texture. So it is the most showy part of the blossom. Fig. 5 shows the corolla of the Morning-Glory whole. Fig. 7 is the same, split down and spread open to show

15. The Stamens. These in this flower grow fast to the bottom of the corolla. There are five stamens in the Morning-Glory. Each stamen consists of two parts, namely, a *Filament* and an *Anther.* The *Filament* is the stalk; the *Anther* is a little case, or hollow body, borne on the top of the filament. It is filled with a powdery matter, called *Pollen.* Fig. 9 shows a separate stamen on a larger scale: *f,* the filament; *a,* the anther, out of which pollen is falling from a slit or long opening down each side.

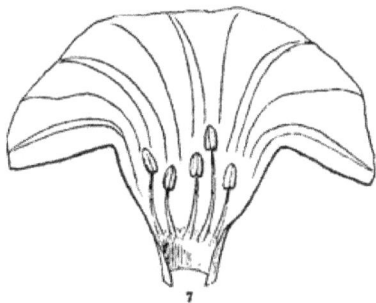

16. **The Pistils** are the bodies in which the seeds are formed. They belong in the centre of the flower. The Morning-Glory has only one pistil: this is shown, enlarged, in Fig. 8. The Rose and the Buttercup have a great many. A pistil has three parts. At the bottom is the *Ovary*, which becomes the seed-vessel. This is prolonged upwards into a slender body, called the *Style*. And this bears a moist, generally somewhat enlarged portion, with a naked roughish surface (not having any skin, like the rest), called the *Stigma*. Upon this stigma some of the pollen, or powder from the anthers, falls and sticks fast. And this somehow enables the pistils to ripen seeds that will grow.

17. Let us now look at a stamen and a pistil from one of the flowers of a Lily (like those shown on a reduced scale in Figures 1 and 2, on the first page), where all the parts are on a larger scale. Here is a *Stamen* (Fig. 9), with its stalk or *Filament, f,* and its *Anther, a,* discharging its yellow dust or *Pollen*. And by its side is the *Pistil* (Fig. 10), with its *Ovary, ov.*; and this tapering into a *Style, st.*; and on the top of this is the *Stigma, stig*. Now cut the ovary through, and it will be found to contain young seeds. Fig. 11 shows the ovary of Fig. 10 cut through lengthwise and magnified by a common hand magnifying-glass. Fig. 12 is the lower part of another one, cut in two crosswise. The young seeds, or more correctly the bodies which are to become seeds, are named *Ovules*. In the Lily these are very numerous. In the Morning-Glory they are few, only six.

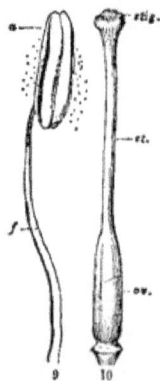

18. These are all the parts of the flower, — all that any flower has. But many flowers have not all these parts. Some have only one flower-cup or one set of blossom-leaves. Lilies appear to have only one set. Some have neither calyx nor corolla; some stamens have no filament, and some pistils have no style: for the style and the filament are not necessary parts, as the anther and the ovary and stigma are. These cases will all be noticed when we come to study flowers more particularly. Meanwhile, please to commit to memory the names of the parts of the flower, Calyx, Corolla, Stamens, and Pistils, and the parts of these also, and learn to distinguish them in all the common blossoms you meet with, until they are as familiar as root, stem, and leaves are to everybody.

19. Notice, also, that the calyx and the corolla, one or both, often consist of separate leaves; as they do in the true Lilies. Each separate piece or leaf of a corolla is called a *Pétal:* and each leaf or piece of a calyx is called a *Sépal.*

20. The corolla, the stamens, and generally the calyx, fall off or wither away after blossoming; while the ovary of the pistil remains, grows larger, and becomes

21. **The Fruit.** So that the fruit is the ripened ovary. It may be a berry, a stone-fruit, a nut, a grain, or a pod. The fruit of the Lily and also of the Morning-Glory is a pod. Here is the pod or fruit of the Morning-Glory (Fig. 4, *fr.* and Fig. 13), with the calyx remaining beneath, and the remains of the bottom of the style resting on its summit. And Fig. 14 shows the same pod, fully ripe and dry, and splitting into three pieces that the seeds may fall out. This pod has three cavities (called *Cells*) in it; and in each cell two pretty large seeds. Lily-pods have three cells, as we may see in the ovary in the flower (Fig. 12), and many seeds in each.

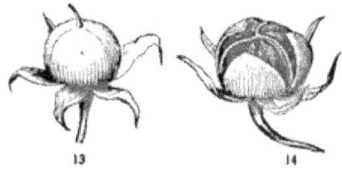

22. **Seeds.** These are the bodies produced by the ripened pistil, from which new plants may spring. Here (Fig. 15) is a seed of Morning-Glory, a little enlarged. Also two seeds cut through lengthwise in two different directions, and viewed with a magnifying-glass, to show what is inside (Fig 16, 17). The part of the seed that grows is

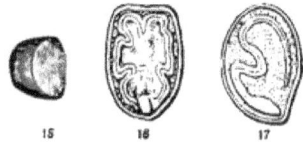

23. **The Embryo, or Germ.** This is a little plantlet ready formed in the seed. In the Morning-Glory it is pretty large, and may readily be got out whole from a fresh seed, or from a dried one after soaking it well in hot water. In Fig. 16 it is shown whole and flatwise in the seed, where it is a good deal crumpled up to save room. In Fig. 17, merely the thickness of the embryo is seen, edgewise, in the seed, surrounded by the pulpy matter, which is intended to nourish it when it begins to grow. In Fig. 18, the embryo is shown taken out whole, and spread out flat. In Fig. 19, its two little leaves are separated, and we plainly see what it consists of. It is a pair of tiny leaves on the summit of a little stem. The leaves (Fig. 19, *c, c*) are named *Seed-leaves* or *Cotylèdons;* the little stem or *stemlet* is named the *Radicle, r.*

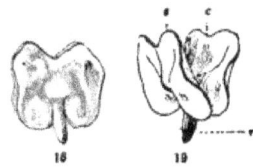

Analysis of the Section.

1.* Plants consist of two kinds of Organs : those of Vegetation ; what they are: 2. those of Reproduction; what they are, what their use.

4. The Root; what it is; rootlets. 5. The Stem; what it is, what it bears. 6. Leaves. 7. The Plant in Vegetation; action of the root, stem, and leaves: they change earth, air, and water into nourishment, and use this nourishment in growing. 8. Shrub or tree like an herb, only more extended.

9. The plant reproduces itself, by seed; blossoming. 10. Object of flowers, fruit, seed: all intended for producing the germ or embryo; what this is.

11. Flowers, why particularly interesting to the botanist. 12. What the parts of a flower are; Floral Envelopes; Essential Organs, why so called.

13. Calyx. 14. Corolla. 15. Stamens; what they consist of; Filament; Anther; Pollen. 16. Pistils; how situated; parts of a pistil; Ovary, Style, Stigma; its use. 17. Stamens and pistil shown in another flower, and the parts explained: Ovules, what they are. 18. All these parts not always present; what ones often wanting. 19. Leaves of a corolla, called Petals; of a calyx, Sepals. 20. What becomes of the parts of a blossom.

21. Fruit, what it is, what it contains. 22. Seeds, what they are, what the part is that grows. 23. Embryo or Germ; what it consists of: Cotyledons or Seed-leaves; Radicle or Stemlet.

SECTION II. — **How Plants grow from the Seed.**

24. **Illustrated by the Morning-Glory.** We now know what all the parts of a plant are ; that a plant, after growing or vegetating awhile, blossoms ; that flowers give rise to fruit ; that the fruit contains one or more seeds ; and that the essential part of a seed is the embryo or germ of a new plant. To produce, protect, and nourish this germ, is the object of the flower, the fruit, and the seed. The object of the embryo is to grow and become a new plant. How it grows, is what we have now to learn.

25. **Life in a Seed.** But first let us notice that it does not generally grow at once. Although alive, a seed may for a long while show no signs of life, and feel neither the summer's heat nor the winter's cold. Still it lives on where it falls, in this slumbering way, until the next spring in most plants, or sometimes until the spring after that, before it begins to grow. There is a great difference in this respect in different seeds. Those of Red Maple ripen in the spring, and start about the middle of the summer. Those of Sugar Maple ripen in the fall, and lie quiet until the next spring. When gathered and laid up in a dry place, many seeds will keep alive for two, three, or several years; and in this state plants may be safely transported

---

* The numbers are those of the paragraphs.

all around the world.   How long seeds will live is uncertain.   The stories of seeds growing which have been preserved for two or more thousand years with Egyptian mummies, are not to be believed.   But it is well known that Sensitive Plants have been raised from seeds over sixty years old.   Few kinds of seeds will grow after keeping them for five or six years; many refuse to grow after the second year; and some will not grow at all unless allowed to fall at once to the ground.   There is no way of telling whether the germ of a seed is alive or not, except by trying whether it will grow, that is, will *germinate*.

26. **Germination and Early Growth.**   *Germination* is the sprouting of a plant from the seed.   Having just illustrated the parts of a plant by the Morning-Glory, from the root up to the seed and the embryo in the seed, we may take this same plant as an example to show how a plant grows from the seed.   If we plant some of the seeds in a flower-pot, covering them lightly with soil, water them, and give them warmth, or if in spring we watch those which sowed themselves naturally in the garden the year before, and are now moistened by showers and warmed by sunshine, we shall soon see how they grow.   And what we learn from this one kind of plant will be true of all ordinary plants, but with some differences in the circumstances, according to the kind.

27. The seed first imbibes some moisture through its coats, swells a little, and, as it feels the warmth, the embryo gradually wakes from its long and deep sleep, and stretches itself, as it were.   That is, the tiny stem of the embryo lengthens, and its end bursts through the coats of the seed; at the same time, the two leaves it bears grow larger, straighten themselves, and so throw off the seed-coats as a loose husk; this allows the seed-leaves to spread out, as leaves naturally do, and so the seedling plantlet stands revealed.   Observe the whole for yourselves, if possible, and compare with these figures.   Fig. 19 is repeated from p. 9, and represents the embryo taken out of the seed, straightened, enlarged, and the two leaves a little opened.   Fig. 16 and 17 show how the embryo lies snugly packed away in the seed.   Fig. 20 shows it coming up, the seed-leaves above just throwing off the coats or husk of the seed.   Fig. 21 is the same, a little later and larger, with the seed-leaves spread out in the air above, and a root well formed beneath.   And Fig. 22 is the same a little later still.

28. At the very beginning of its growth, the end of the little stem which first comes out of the seed turns downward and points into the earth.   From it the root is formed, which continues downwards, branching as it grows, and burying itself

more and more in the soil. The other end of the stem always turns upwards, and, as the whole lengthens, the seed-leaves are brought up out of the ground, so that they expand in the light and air, — which is the proper place for leaves, as the dark and damp soil is for the root.

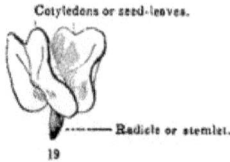

Cotyledons or seed-leaves.

Radicle or stemlet.

19

29. What makes the root always grow downwards into the ground, and the stem turn upwards, so as to rise out of it, we no more know, than we know why newly-hatched ducklings take to the water at once, while chickens avoid it, although hatched under the same fowl and treated just alike. But the fact is always so.

And although we know not *how*, the *why* is evident enough; for the root is thereby at once placed in the soil, from which it has to absorb moisture and other things, and the leaves appear in the air and the light, where they are to do their work.

30. Notice how early the seedling plant is complete, that is, becomes a real vegetable, with all its parts, small as the whole thing is (Fig. 21). For it already possesses a root, to connect it with the ground and draw up what it needs from that; a stem, to elevate the foliage into the light and air; and leaves, to take in what it gets directly from the

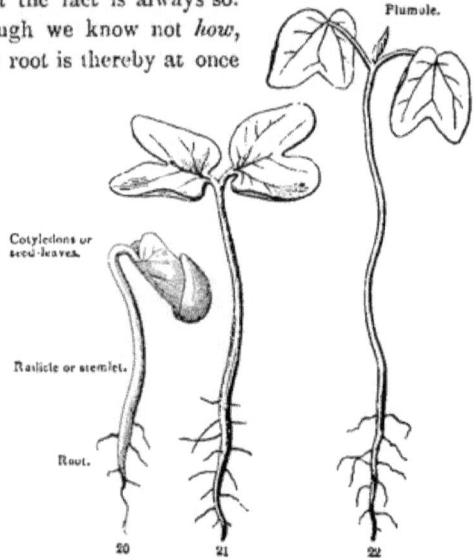

Plumule.

Cotyledons or seed-leaves.

Radicle or stemlet.

Root.

20   21   22

air, and to digest the whole in the light (as explained in the last section, Par. 7). That is, it already has all the Organs of Vegetation (Par. 1), all that any plant has before blossoming, so that the little seedling can now take care of itself, and live — just as any larger plant lives — upon the soil and the air. And all it has to do in order to become a full-grown plant, like Fig. 4, is to increase the size of its organs, and to produce more of them; namely, more stem with more leaves above, and more roots below. We have only to watch our seedling plantlets a week or two longer, and we shall see how this is done.

31. The root keeps on growing under ground, and sending off more and more small branches or *rootlets*, each one adding something to the amount of absorbing surface in contact with the moist soil. The little stem likewise lengthens upwards, and the pair of leaves on its summit grow larger. But these soon get their full growth; and we do not yet see, perhaps, where more are to come from. But now a little bud, called the *Plumule*, appears on the top of the stem (Fig. 22), just between the stalks of the two seed-leaves; it enlarges and unfolds into a leaf; this soon is raised upon a new piece of stem, which carries up the leaf, just as the pair of seed-leaves were raised by the lengthening of the radicle or first joint of stem in the seed. Then another leaf appears on the summit of this joint of stem, and is raised upon its own joint of stem, and so on. Fig. 23 shows the same plant as Fig. 22 (leaving out the root and the lower part of the stem), at a later stage; *c, c,* are the seed-leaves; *l* is the next leaf, which came from the plumule of Fig. 22, now well raised on the second joint of stem; and *l'* is the next, still very small and just unfolding. And so the plant grows on, the whole summer long, producing leaf after leaf, one by one, and raising each on its own joint of stem, arising from the summit of the next below; — as we see in Fig. 4, at the beginning of the chapter, where many joints of stem have grown in this way (the first with a pair of leaves, the rest with one apiece), and still there are some unfolding ones at the slender young summit.

32. **How the Seedling is nourished at the Beginning.** *Growth* requires *food*, in plants as well as in animals. To grow into a plant, the embryo in a seed must be fed with vegetable matter, or with something out of which vegetable matter can be made. When a plant has established itself, — that is, has sent down its roots into the soil, and spread out some leaves in the air, — it is then able to change mineral matter (viz. earth, air, and water) which it takes in, into vegetable matter, and so to live and grow independently. But at the beginning, before its organs are developed and established in their proper places, the forming plant must be supplied by ready-made vegetable matter, furnished by the mother plant. On this supply the embryo germinating from the seed feeds and grows, — just as the new-

born animal does upon the mother's milk, or as the chick developing in the egg does upon the prepared nourishment the parent had laid up for the purpose in the yolk.

33. Tear open a fresh Morning-Glory seed, or cut a dried one in two, as in Fig. 17, and this supply will be seen, in the form of a rich and sweetish jelly-like matter, packed away with the embryo, and filling all the spaces between its folds. This is called the *Albumen* of the seed (that being the Latin name of the white of an egg) ; and this is what the embryo feeds upon, and what enables its little stemlet (Fig. 19, *r*) to grow, and form its root downwards, and carry up and expand its seed-leaves (*c, c*) in the air, and so become at once a plantlet (Fig. 21), with root, stem, and leaves, able to take care of itself, just as a chicken does when it escapes from the shell.

34. This moist nourishing jelly would not keep long in that state. So, when the seed ripens and dries, it hardens into a substance like thin dried glue or gum, which will keep for any length of time. And whenever the seed is sown, and absorbs moisture, this matter softens into a jelly again, or gradually liquefies, and the seed-leaves crumpled up among it drink it in at every pore. A portion is consumed in their growth, while the rest is carried into the growing stemlet, thence into the root forming at one end of it, and into the bud (or *plumule*, Fig. 22) which soon appears at the other end of it, — supplying the materials for their growth.

35. Notice the same thing in Wheat, Oats, or Indian Corn. The last is the best example, because the grain is so large that all the parts may be clearly seen without magnifying. The abundant *milk* or soft and rich pulp of green corn is the same as the jelly in the seed of the Morning-Glory ; namely, it is the *albumen* of the seed, provided for the embryo (the chit or germ) to feed upon when growth begins. See Figures 44, 45, &c. This nourishing food (as we well know it to be) was produced by the mother-plant during the summer, was accumulated in the stalk at flowering-time, in the form of sugar, or syrup, was conveyed into the flowers and forming seeds ; a part was used to form the germ or embryo, and the rest was stored up with it in the seed, to serve for its growth into a plantlet the next spring. That it may keep through the winter, or longer, the sweet milk is changed into a starchy pulp, which hardens as the grain ripens into the firm and dry mealy part (or *albumen*), which here makes the principal bulk of the seed. But when sown, this meal softens and is slowly changed back into sugar again. And this, dissolved in the water the seed takes in, makes a sweet sap, which the

embryo imbibes and feeds on as it sprouts. That the meal or starch of the grain is actually changed into sugar at this time is clearly shown by malting, which is merely causing heaps of grain to sprout a little, and then destroying the life of the embryo by dry heat; when the grain (now malt) is found to be sweet, and to contain much sugar.

36. The nourishment which the mother-plant provides in the seed is not always stored up outside of the embryo. In many cases it is deposited *in the embryo* itself, most commonly in the seed-leaves. Then the seed consists of nothing but the embryo within its coats. Maple-seeds are of this sort. Fig. 24 represents a seed of Red Maple in the lower part of the winged seed-vessel, which is cut away so as to show it in its place. Fig. 25 is the seed a little magnified, and with the coats cut away, bringing to view its embryo coiled up within and filling the seed completely. Fig. 26 is the embryo taken out, and a little unfolded; below is the radicle or stemlet; above are the two seed-leaves partly crumpled together. Fig. 27 is the embryo when it has straightened itself out, thrown off the seed-coats, and begun to grow. Here the seed-leaves are rather thick when they first unfold; this is on account of the nourishing matter which was contained in their fabric, and which is used mainly for the earliest growth of the radicle or stemlet, and for the root formed at its lower end, as we see in the next figure (Fig. 28: *a*, the radicle or stemlet of the embryo; *b, b*, the two seed-leaves; *c*, the root). By this time the little stock of nourishment is exhausted. But the plant, having already a root in the soil and a pair of leaves in the air, is able to shift for itself, to take in air, water, &c., and by the aid of sunshine on its foliage to make the nourishment for its future growth. In a week or two it will have made enough to enable the next step to be taken. Then a little bud appears at the upper end of the stemlet, between the two seed-leaves, and soon it shows the rudiments of a new pair of leaves (Fig. 28, *d*); a new joint of stem forms to support them (Fig. 29); this lengthens just as the stemlet of the embryo did, and so the plantlet gets a second pair of leaves, raised on a second joint of stem

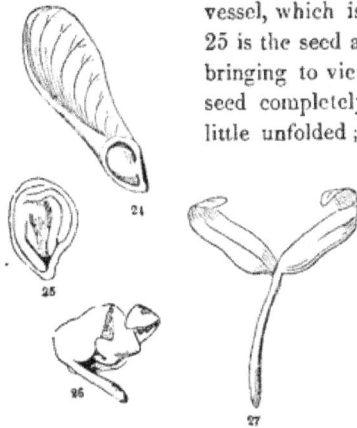

2

springing from the top of the first (Fig. 30). Meanwhile the root has grown deeper into the soil, and sent out branches. Having now more roots below, and, above, a pair of leaves besides the seed-leaves to work with, the seedling plantlet

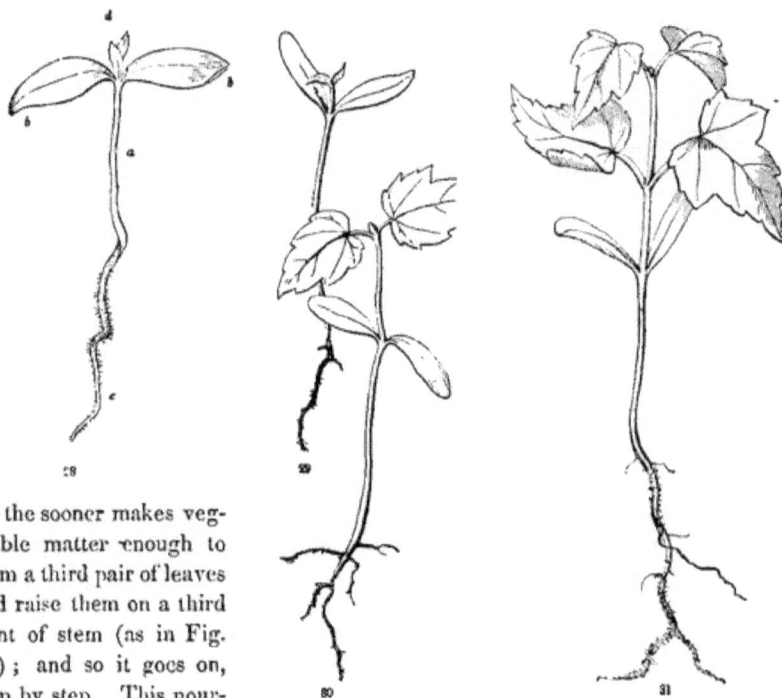

all the sooner makes vegetable matter enough to form a third pair of leaves and raise them on a third joint of stem (as in Fig. 31); and so it goes on, step by step. This nourishment in the embryo of the Red-Maple seed was a few weeks before in the trunk of the mother tree, as a sweet sap, that is, as *Maple-sugar*.

37. **Variations of the Plan of Growth.** In the Morning-Glory, after the pair of seed-leaves, only one leaf is found upon each joint of stem (see Fig. 23 and 4). In the Maple there is a pair of leaves to every joint of stem, as long as it grows. In the Morning-Glory the food in the seed, for the growth to begin with, was stored up *outside of the embryo;* in the Maple it was stored up *in it,* that is, in its seed-leaves. The plan is evidently the same in both; but there are differ-

ences in the particulars. While the same kind of plant always grows in exactly the same way, different kinds differ almost as much at the beginning as they do afterwards. The great variety which we observe among the herbs and shrubs and trees around us, — in foliage, flower, fruit, and everything, — gives to vegetation one of its greatest charms. We should soon tire of plants or flowers made all after one exact pattern, however beautiful. We enjoy variety. But the botanist finds a higher interest in all these differences than any one else, because he discerns one simple plan running through all this diversity, and everywhere repeated in different forms. He sees that in every plant there is root growing downwards, connecting the vegetable with the soil; stem rising into the light and air, and bearing leaves at regular places, and then blossoms, and that the parts of one kind of blossom answer to those of another, only differing in shape; and he delights in observing how the tens of thousands of kinds of plants all harmonize with each other, like the parts of concerted music, — plainly showing that they were all contrived, as parts of one system, by one Divine Mind.

38. So in the beginning, in the growth of plants from the seed, although the general plan is the same in all, the variations are many and great. The plan is well shown in the two seedling plants which have served for illustration, namely, the Morning-Glory and the Maple. Let us now notice some of the variations, as exhibited in a few very common plants. A great deal may be learned from the commonest plants, if we will only open our eyes to see them, and "consider how they grow," and why they differ in the way they do. Take, for instance,

39. **The Bean.** Soak a bean in warm water (if a fresh one is not to be had) and remove the coats. The whole kernel consists of an embryo, as seen in Fig. 32. And almost the whole bulk of this embryo consists of two thick pieces, c, c, which are the cotyledons or seed-leaves. We may make out the plan of the whole thing better by spreading these thick seed-leaves wide open, as in Fig. 33. Here the two thick seed-leaves are seen from the inside, c, c; they are connected with the upper end of a stemlet, which is the radicle, r; and above this already shows the bud or plumule, p.

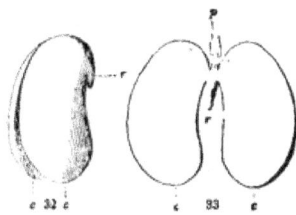

40. So the embryo of the Bean is the same in plan as that of the Maple (Fig. 27), only the stemlet is much shorter in proportion, and the seed-leaves very much larger and thicker. What is the reason of this difference?

41. The seed-leaves of the Bean are thickened by having so much nourishment stored up in them, so much of it that they make good food for men. And the object of this large supply is that the plant may grow more strongly and rapidly

from the seed. It need not and it does not wait, as the Maple and the Morning-Glory do, slowly to make the second pair of leaves; but is able to develop these at once. Accordingly, the rudiments of these next leaves may be seen in the seed before growth begins, in the form of a little bud (Fig. 33, *p*), ready to grow and unfold as soon as the thick seed-leaves themselves appear above ground (Fig. 34), and soon making the first real foliage (Fig. 35). For the seed-leaves of the Bean are themselves so thick and ungainly, that, although they turn green, they hardly serve for foliage. But, having given up their great stock of nourishment to the forming root and new leaves, and enabled these to grow much stronger and faster than they otherwise could, they wither and fall off. It is nearly the same in

42. The Cherry, Almond, &c. Fig. 36 is an Almond taken out of the shell, soaked a little, and the thin seed-coat removed. The whole is an embryo, consisting of a pair of large and thick seed-leaves, loaded with sweet nourishment. These are borne on a very short radicle, or stemlet, which is seen at the lower end. Pull off one of the seed-leaves, as in Fig. 37, and you may see the plumule or little bud, *p*, ready to develop leaves and stem upwards, while the other end of the radicle grows downward and makes the root; the rich store of nourishment in the seed-

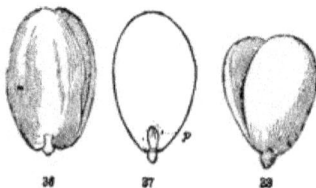

leaves supplying abundant materials for the growth. A cherry-seed is just like an almond, only on a smaller scale. Fig. 38 is the embryo of a Cherry, with the very thick seed-leaves a little separated. Fig. 39 is the same developed into a young plantlet. Fed by the abundant nourishment in the seed-leaves, it shoots up its stem and unfolds three or four leaves before the Maple (Fig. 28, 29) or the Morning-Glory (Fig. 20 – 22) would have made any. It is the same in the Chestnut and the Beech. In these, as in the Cherry and the Bean, the thick seed-leaves, which make the whole kernel, come up, turn green, and become thinner as they give up their load of nourishment to the growing parts; they evidently try to become useful green leaves; but having been used for holding nourishment, they remain too thick and clumsy for foliage, and they soon die or fall off. But in

43. The Horsechestnut, the Acorn, and the Pea, the seed-leaves are so very thick, and so heavily loaded, that they never undertake to serve any other purpose than that of feeding the other parts as they grow. So they remain in the shell or husk; and, as they are not to rise out of the ground, there is no need for their stemlet, or radicle, to lengthen, except enough to get

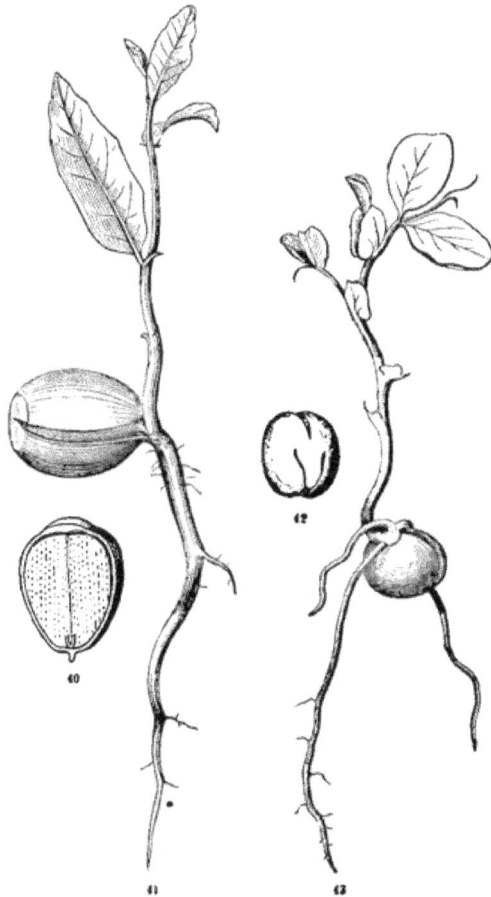

out of the seed, and let the root form from the lower end of it, while the plumule develops from its upper end directly into a strong leafy stem. Fig. 40 is an acorn cut through lengthwise. The whole kernel consists of a pair of very thick seed-leaves, loaded with starch, &c., and completely enclosing the very small and short stemlet, or radicle, seen at the bottom. Fig. 41 is the acorn with the seedling Oak growing from it; the seed-leaves remaining in the shell, but feeding the strong root which grows downwards and the stem which shoots so vigorously upwards.

44. Acorns and horsechestnuts may not always be found germinating; but in the Pea we have a familiar case of this way of growing, which may be observed at any season by planting a few peas. Fig. 42 is a pea with the seed-coat taken off, after soaking. Here the seed-leaves are so thick that the pair makes a little ball; and the stout radicle or stemlet appears on the side turned to the eye. Fig. 43 shows the plantlet growing. The whole seed remains in the soil; the plumule, well nourished by the great stock of food in the buried seed-leaves, alone rises out of the ground as a strong shoot, bearing an imperfect scale-like leaf upon each of its earlier joints, and then producing the real leaves of the plant, while the radicle at the same time, without lengthening itself, sends down three or four roots at once. So the whole plant is quickly established, and all the early growth is made out of food provided for it the year before by the mother plant, and stored up in the seed. One more illustration we may take from

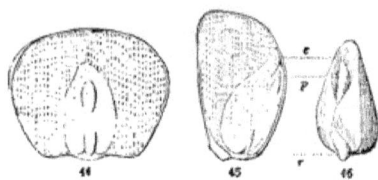

45. **Indian Corn.** Here the food provided for the early growth is laid up partly in the embryo, but mostly around it. Fig. 44 is a grain cut through flat-wise; Fig. 45, another cut through the middle across its thickness; and Fig. 46, the embryo, or germ, of another grain, taken out whole, — which may readily be done in green corn, or in an old grain after soaking it for some time in warm water. The separate embryo is placed to match that which is seen, divided, in the seed; r is the radicle; p, the plumule; and c, the seed-leaf or cotyledon, which in this plant is single; while in all the foregoing there was a pair of seed-leaves. The greater part of the grain is the meal, or albumen, the stock of nourishment outside of the embryo. In germinating, this meal is slowly changed

into sugar, and dissolved in the water which is absorbed from the ground ; the cotyledon imbibes this, and sends it into the radicle, r, to make the root, and into the
plumule, p, enabling it to develop the set of leaves,
wrapped up one within another, of which it consists,
and expand them one after another in the air. Fig.
47 shows a sprouting grain, sending down its first
root, and sending up the plumule still rolled together.
Fig. 48 is the same, more advanced, having made a
whole cluster of roots, and unfolded two or three
leaves. Nourished abundantly as it is, both by the
maternal stock in the grain, and by what these roots
and leaves obtain and prepare from the soil and the
air, the young corn gets a good start, is ready to avail
itself of the summer's heat, to complete its vegetation, to blossom, and to make and lay up the great
amount of nourishment which we gather in the crop.

46. **The Onion.** The cotyledon in Indian Corn, and
most other plants which have only one, stays under
ground. In the Onion it comes up and makes the
first leaf, — a slender, thread-shaped one, — and indeed it carries up the light seed on its summit. In
Indian Corn, all the early joints of stem remain so
short as not to be seen ; although later it makes long
joints, carrying up the upper leaves to some distance
from one another. In the Onion, on the contrary, the
stem never lengthens at all, but remains as a thin
plate, broader than it is long, with the roots springing from one side of it and the
sheathing bases of the leaves covering it on the other.

47. **Number of Cotyledons or Seed-Leaves.** Indian Corn (Fig. 46) and all such
kinds of grain-plants, the Onion, Lilies, and the like, have only one seed-leaf or
cotyledon to their embryo ; therefore they are called MONOCOTYLEDONOUS PLANTS,
and the embryo is called *monocotylédonous*, — a long word, meaning "with one
cotyledon."

48. The embryo of the Morning-Glory (Fig. 19), of the Maple (Fig. 27),
Bean (Fig. 32 – 34), Almond, Peach, and Cherry (Fig. 36 – 38), Oak (Fig. 40),

Pea (Fig. 42), and of all such plants, is *dicotylédonous*, that is, has a pair of cotyledons, or seed-leaves, which is what the word means. Therefore all such plants are called DICOTYLEDONOUS PLANTS.

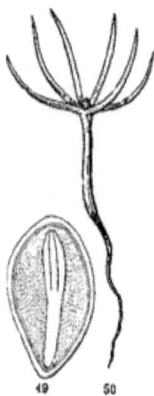

49. Pine-trees, and plants like them, generally have more than two cotyledons, in a circle ; so their embryo is said to be *poly-cotylédonous;* meaning " with several or many cotyledons." Fig. 49 is a magnified view of a Pine-seed, divided lengthwise, and showing the long and straight embryo lying in the middle of the albumen. The slender lower part is the radicle or stemlet; the upper part is a cluster of cotyledons or seed-leaves, in a close bundle; three of them can be seen as it lies, and there are as many more behind. Fig. 50 is this embryo as it comes up from the seed, its cotyledons (six in number) expanding at once into a circle of slender, needle-shaped leaves.

50. It is a pity these three words are so long; for the pupil should fix them thoroughly in his memory; because these differences in the embryo, or plantlet in the seed, run through the whole life of the plant, and show themselves in many other differences which very strikingly distinguish one class of plants from another. Let it be remembered, therefore, that

Monocotyledonous Plants, or *Monocotyledons*, are those which have only one cotyledon or seed-leaf to their embryo.

Dicotyledonous Plants, or *Dicotyledons*, are those which have a pair of cotyledons or seed-leaves to their embryo.

Polycotyledonous Plants, or *Polycotyledons*, are those which have more than one pair of cotyledous or seed-leaves to their embryo.

Analysis of the Section.

24. Flowers produce Fruit; this, the Seed; of this the essential part is the Embryo which grows. 25. It is alive; but lies dormant awhile. How long seeds may live.

26. Germination, the beginning of growth; what is needful for it. 27. What takes place, illustrated from the Morning-Glory. 28. How the stemlet grows by lengthening, and carries up the seed-leaves: how the root is formed and grows downwards. 29. Instinct of each part to turn in its proper direction; and why. 30. The little seedling a complete plant in miniature; its parts. 31. How it goes on to grow: growth of the root; rootlets; of the stem. The Plumule or Bud. Development of the stem piece by piece, each with its leaf.

32. How the seedling is nourished at the beginning. Growth requires food. 33. How this is supplied by a deposit in the seed; Albumen. 34. It is kept in a solid form until the embryo starts, and is

then dissolved, turned into sugar, &c., and feeds the plantlet. 35. This illustrated in Wheat and Indian Corn. 36. Or else the same nourishment is deposited in the embryo itself, in its seed-leaves; illustrated by the Maple. 37, 38. Variations of the same plan of growth in different plants. The Maple compared with the Morning-Glory. 39–45. A great abundance of food stored up in the embryo causes a rapid and strong growth; illustrated by the Bean; 42. by the Cherry, Almond, &c.; 43, 44. by the Horsechestnut, Acorn, Pea, &c.; in these the seed-leaves do not come up in germinating; why. 45. In Indian Corn; the stock of food partly in the strong embryo, partly outside of it. 46. The Onion; its seed-leaf lengthens and comes up, but the stem never lengthens at all.

47. Number of cotyledons or seed-leaves in different kinds of plants; Monocotyledonous. 48. Dicotyledonous ; Polycotyledonous. 50. These differences always accompany other differences in the plant; Monocotyledonous, Dicotyledonous, and Polycotyledonous Plants.

SECTION III. — How Plants grow Year after Year.

51. **They Grow on as they Began.** The seedling has all the organs that any plant has, — even the largest and oldest, — excepting what belongs to blossoms : it has all it needs for its life and growth, that is, for vegetation. It has only to go on and produce more of what it already has, — more roots beneath to draw up more moisture from the soil, and more stem above, bearing more leaves, exposing a larger surface to the light and air, in which to digest what is taken in from the soil and the air, and turn it into real nourishment, that is, into the stuff which vegetables are made of. So, as fast as a young plant makes new vegetable material, it uses it for its growth; it adds to its root below, and to its stem above, and unfolds a new leaf or pair of leaves on every joint. Each joint of stem soon gets its full length, and its leaf or pair of leaves the full size; and now, instead of growing, they work, or prepare nourishment, for the growth of the younger parts forming above.

52. **Simple Stems.** In this way, piece by piece, the stem is carried up higher and higher, and its leaves increased in number ; and the more it grows, the more it is able to grow, — as we see in a young seedling, beginning feebly and growing slowly for a while, but pushing on more and more vigorously in proportion to the number of leaves and roots it has produced. In this way, by developing joint after joint, each from the summit of its predecessor, a *Simple Stem* is made. Many plants make only simple stems, at least until they blossom, or for the first year. The Lilies, figured on the first page, and corn-stalks, are of this kind. Fig. 51 is a sort of diagram of the simple stem of Indian Corn, divided into its component pieces, to show how it consists of a set of similar growths, each from the summit

of the preceding one. There are old trees even, which consist of a simple, unbranched stem. Palm-trees, such as our Southern Palmetto (Fig. 79) are of this
kind. But more commonly, as stems grow they multiply themselves by forming

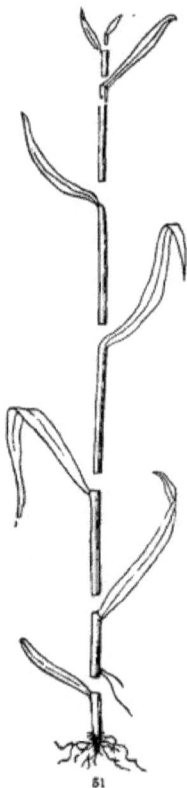

53. **Branches,** or *side-shoots.* These are formed both by
roots and by stems. Roots generally branch much sooner than
stems do. See Fig. 4, 20, 30, &c.

54. Roots send off their branches from any part of the
main root, or start from any part of a stem lying on or in the
soil; and they have no particular arrangement.

55. But the branches of stems spring only from particular
places, and are arranged on a regular plan. They arise from
the *Axil* of a leaf and nowhere else, except in some few peculiar cases. The *axil* (from a Latin word meaning the
armpit) of a leaf is the hollow or angle, on the upper side,
where the leaf is attached to the stem. As branches come
only from the axils of leaves, and as leaves have a perfectly
regular and uniform arrangement in each particular plant, the
places where branches will appear are fixed beforehand by the
places of the leaves, and they must follow their arrangement.
In the axils, commonly one in each, branches first appear in
the form of

56. **Buds.** A *Bud* is an undeveloped stem or branch. If
large enough to have its parts distinguishable, these are seen
to be undeveloped or forming leaves: and large buds which
are to stand over winter are generally covered with protecting scales, — a kind of dry, diminished leaves.

57. **Terminal Bud.** So the *plumule* or first shoot of the
embryo (see Fig. 22, &c.) is a bud. But this first bud makes
the main stem, and its growth, week after week, or year after
year, carries on the main stem. Palms (as Fig. 79) grow in this way, by this bud
only. Being always on the end of the stem, that is, terminating the stem, it is
called the *Terminal Bud.*

58. **Axillary Buds.** But the buds which are to form branches appear on the
sides of the stem; and since they are situated in the axils of the leaves, as just ex-

plained (55), they are named *Axillary Buds*. (See Fig. 52, 53.) These buds grow into branches, just as the first or terminal bud of the seedling grows to make the main stem.

59. **The Arrangement of Branches**, therefore, follows that of the axillary buds, and this that of the leaves. Now leaves are placed on the stem in two principal ways; they are either *alternate* or *opposite*. They are *alternate* when they follow one after another, there being only one to each joint of the stem, as in Morning-Glory (Fig. 4, all after the seed-leaves), and in the Linden or Basswood (Fig. 52), as well as the greater part of trees or plants. They are *opposite* when there are two leaves upon each joint of stem, as in Horsechestnut, Lilac, and Maple (Fig. 31, 53); one leaf in such cases being always exactly on the opposite side of the stem from its fellow. Now in the axil of almost every leaf of these trees a bud is soon formed, and in general plainly shows itself before summer is over. In Fig. 52, *a, a, a, a,* are the *axillary buds* on a twig of Basswood,—they are *alternate*, like the leaves,—and *t* is the *terminal bud*. Fig. 53, a twig of Red Maple, has its axillary buds *opposite*, like the leaves; and on the very summit is the *terminal bud*. Next spring or sooner, the former grow into *alternate branches*; the latter grow into *opposite branches*. These branches in their turn form buds in the axils of their leaves, to grow in time into a new generation of similar branches, and so on, year after year. So the reason is plain why the branching or spray of one tree or bush differs from that of another, each having its own plan, depending upon the way the leaves are arranged on the stem.

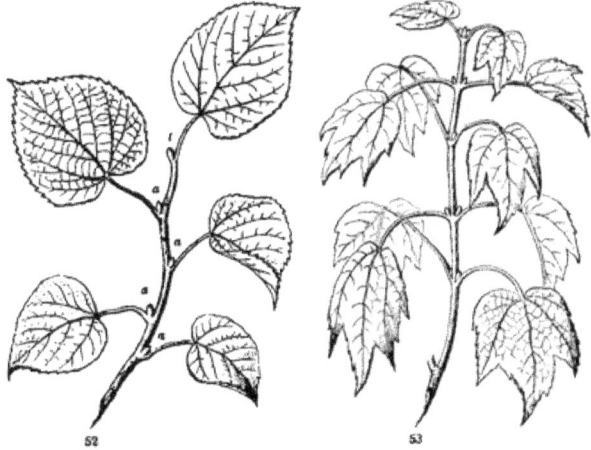

60. The spray (or *ramification*) of trees and shrubs is more noticeable in winter,

when most leaves have fallen.   Even then we can tell how the leaves were placed, as well as in summer.   We have only to notice the *leaf-scars :* for each fallen leaf has left a scar to mark where its stalk separated from the stem.   And in most cases the bud above each scar is now apparent or conspicuous, ready to grow into branches in the spring, and showing plainly the arrangement which these are to have.   Here, for instance, is a last year's shoot of Horsechestnut (Fig. 54), with a large terminal bud on its summit, and with very conspicuous leaf-scars, *l s ;* and just above each is an axillary bud, *b.*   Here the leaves were opposite each other; so the buds are also, and so will the branches be, unless one of the buds on each joint should fail.   Fig. 55 is a similar shoot of a Hickory, with its leaf-scars (*l s*) and axillary buds (*b*) *alternate,* that is, single on the joints and one after another on different sides of the stem ; and these buds when they grow will make alternate branches.

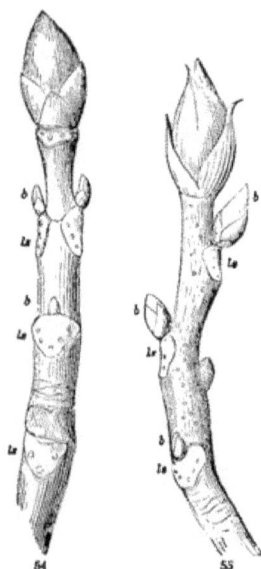

61. The branching would be more regular than it is, if all the buds grew.   But there is not room for all ; so only the stronger ones grow.   The rest stand ready to take their place, if those happen to be killed. Sometimes there are more buds than one from the same axil.   There are three placed side by side on those shoots of Red Maple which are going to blossom. There are several in a row, one above another, on some shoots of Tartarean Honeysuckle.

62. The appearance of plants, the amount of their branching, and the way in which they continue to grow, depend very much upon their character and duration.

63. **The Duration of Plants** of different kinds varies greatly.   Some live only for a few months or a few weeks; others may endure for more than a thousand years. The most familiar division of plants according to their duration and character is into *Herbs, Shrubs,* and *Trees.*

64. **Herbs** are plants of soft texture, having little wood in their stems, and in our climate dying down to the ground, or else dying root and all, in or before winter.

65. **Shrubs** are plants with woody stems, which endure and grow year after year, but do not rise to any great height, say to not more than four or five times the

height of a man.   And if they reach this size, it is not as a single main trunk, but by a cluster of stems all starting from the ground.

66. **Trees** are woody plants rising by a trunk to a greater height than shrubs.

67. Herbs are divided, according to their character and duration, into *Annuals*, *Biennials*, and *Perennials*.

68. **Annuals** grow from the seed, blossom, and die all in the same season.   In this climate they generally spring from the seed in spring, and die in the autumn, or sooner if they have done blossoming and have ripened their seed.   Oats, Barley, Mustard, and the common Morning-Glory (Fig. 4) are familiar annuals.   Plants of this kind have *fibrous* roots, i. e. composed of long and slender threads or fibres. Either the whole root is a cluster of such fibres, as in Indian Corn (Fig. 48), Barley (Fig. 56), and all such plants ; or when there is a main or tap root, as in Mustard, the Morning-Glory, &c., this branches off into slender fibres.   It is these fibres, and the slender root-hairs which are found on them, that mainly absorb moisture and other things from the soil ; and the more numerous they are, the more the plant can absorb by its roots.   As fast as nourishment is received and prepared by the roots and leaves, it is expended in new growth, particularly in new stems or branches and new leaves, and finally in flowers, fruit, and seed.   The latter require a great deal of nourishment to bring them to perfection, and give nothing back to the plant in return.   So blossoming and fruiting weaken the plant very

56
Fibrous roots.

much.   Annual plants usually continue to bear flowers, often in great numbers, upon every branch, until they exhaust themselves and die, but not until they have ripened seeds, and stored up in them (as in the mealy part of the grain of Corn, &c., Fig. 44, 45) food enough for a new generation to begin growth with.

69. **Biennials** follow a somewhat different plan.   These are herbs which do not blossom at all the first season, but live over the winter, flower the second year, and then die when they have ripened their seeds.   The Turnip, Carrot, and Parsnip. the Beet, the Radish (Fig. 57), and the **Celandine**, are familiar examples of biennial plants.

70. The mode of life in biennials is to prepare and store up nourishment through the first season, and to expend it the next season in flowering and fruiting.   Accordingly, biennials for the first year are nearly all root and leaves ; these being the organs by which the plant works, and prepares the materials it lives on.   Stem

they must have, in order to bear leaves; for leaves do not grow on roots. But what stem they make is so very short-jointed that it rises hardly any; so that the leaves seem to spring from the top of the root, and all spread out in a cluster close to the ground. As the plant grows, it merely sends out more and more branches of the root into the soil beneath, and adds more leaves to the cluster just above, close to the surface of the warm ground, and well exposed to the light and heat of the sun. Thus consisting of its two working organs only, — root and leaves, — the young biennial sets vigorously to work. The moisture and air which the leaves take in from the atmosphere, and all that the roots take from the soil, are digested or changed into vegetable matter by the foliage while exposed to sunshine; and all that is not wanted by the leaves themselves is generally carried down into the body of the root and stored up there for next year's use. So the biennial root becomes large and heavy, being a storehouse of nourishing matter, which man and animals are glad to use for food. In it, in the form of starch, sugar, mucilage, and in other nourishing and savory products, the plant (expending nothing in flowers or in show) has laid up the avails of its whole summer's work. For what purpose? This plainly appears when the next season's growth begins. Then, fed by this great stock of nourishment, a stem shoots forth rapidly and strongly, divides into branches, bears flowers abundantly, and ripens seeds, almost wholly at the expense of the nourishment accumulated in the root, which is now light, empty, and dead; and so is the whole plant by the time the seeds are ripe.

71. By stopping the flowering, biennials can sometimes be made to live another year, or for many years, or annuals may be made into biennials. So a sort of biennial is made of wheat by sowing it in autumn, or even in the spring and keeping it fed down in summer. But here the nourishment is stored up in the leaves rather than in the roots.

72. The Cabbage is a familiar and more striking example of a biennial in which the store of nourishment, instead of being deposited in the root, is kept in the

leaves and in the short stem or stalk.   These accordingly become thick and nutritious in the Cabbage, just as the root does in the Turnip, or the base of the short stem alone in Kohlrabi, or even the flower-stalks in the Cauliflower; all of which belong to the same family, and exhibit merely different ways of accomplishing the same result.

73. **Perennials** are plants which live on year after year.   Shrubs and trees are of course perennial.   So are many herbs; but in these only a portion generally survives.   Most of our perennial herbs die down to the ground before winter; in many species all but certain separate portions under ground die at the close of the year; but some parts of the stem containing buds are always kept alive to renew the growth for the next season.   And a stock of nourishment to begin the new growth with is also provided.   Sometimes this stock is laid up in the roots, as for instance in the Peony, the Dahlia (Fig. 58), and the Sweet Potato.   Here some thick roots, filled

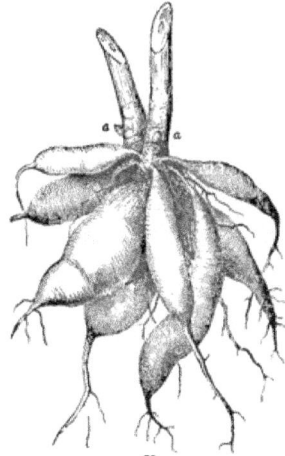
58
Dahlia-roots.

with food made by last year's vegetation, nourish in spring the buds on the base of the stem just above (*a, a*), enabling them to send up stout leafy stems, and send down new roots, in some of which a new stock of food is laid up during summer for the next spring, while the exhausted old ones die off; and so on, from year to year.

74. Sometimes this stock of food is laid up in particular portions of branches of the stem itself, formed under ground, and which contain the buds; as in the Ground Artichoke and the Potato.   Here these parts,

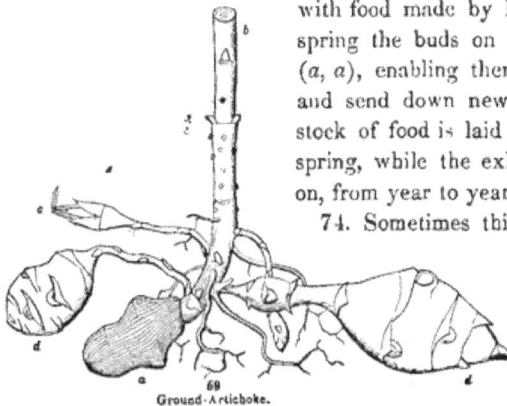
59
Ground-Artichoke.

with their buds, or eyes, are all that live over winter.   These thickened ends of stems are called *Tubers*.   In Fig. 59, *a* is a tuber of last year, now exhausted and withering away, which grew in spring by one of its buds to make the stem (*b*) bearing the foliage of the season.   This sends out some branches under ground, which

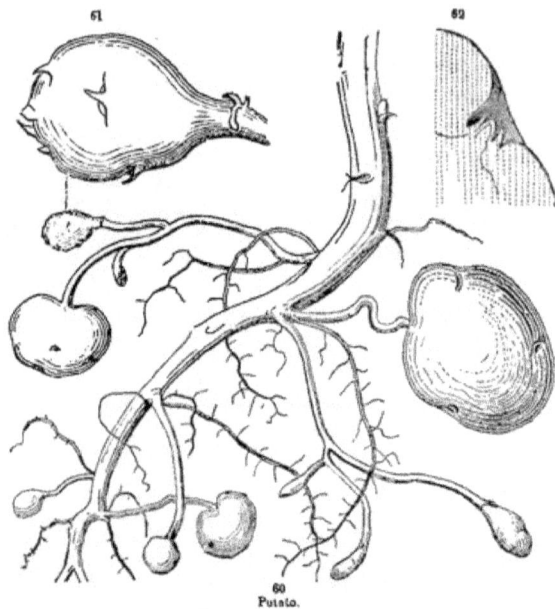

in the course of the season thicken at the end as they receive a stock of nourishment prepared by this year's foliage, and become new tubers (*c*, a forming one ; *d, d,* well-grown tubers of the season), to live over winter and make the next year's growth.

75. Because they live under ground, these tubers are commonly supposed to be roots ; but they are not, as any one may see.   Their eyes are buds ; and the little scales behind the eyes answer to leaves ; while roots bear neither buds nor leaves.   The fibrous roots which grow from these subterranean branches are very different in appearance from under-ground stems, as is plain to see in the

Potato-plant. Fig. 60 shows a few of the real roots, as well as several branches of the stem, with potatoes forming in all stages at their tips. Fig. 61 is one of these forming potatoes magnified, showing a little scale behind each eye which answers to a leaf.   Fig. 62 is a part of a slice through an eye, more magnified, to show that the eye is really a bud, covered with little scales.

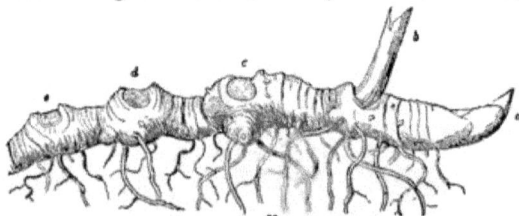

76. In some perennial herbs, prostrate stems or branches under ground are thickened with this store of nourishment for their whole length, making stout *Rootstocks*, as they are called; as in Sweet Flag, Solomon's Seal (Fig. 63), and Iris, or Flower-de-Luce (Fig. 64). These are perennial, and grow on a little way each year, dying off as much behind after a while; and the newer parts every year send out a new set of fibrous roots. The buds which rootstalks produce, and the leaves or the scales they bear, or the scars or rings which mark where the old leaves or scales have fallen or decayed away, all plainly show that rootstocks are forms of stem, and not roots. The large round scars on the root-stock of Solomon's Seal, which give the plant its name, (from their looking like impressions of a seal,) are the places from which the stalk bearing the leaves and flowers of each season has fallen off in autumn. Fig. 63, *a* is the bud at the end, to make the growth above ground next spring; *b* is the bottom of the stalk of this season; *c*, the scar or place from which the stalk of last year fell; *d*, that of the year before; and *e*, that of two years ago.

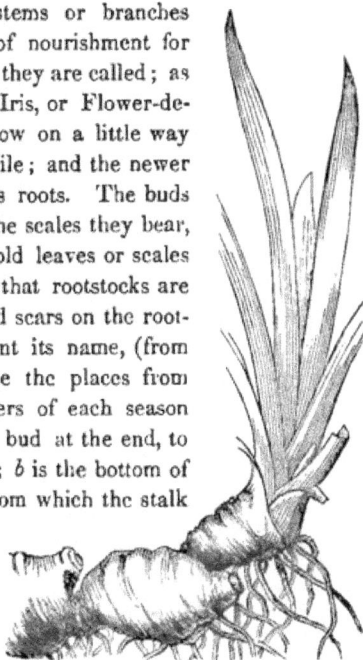

64
Iris.

77. Finally, the nourishment for the next year's growth may be deposited in the leaves themselves. Sometimes it occupies all the leaf, as in the Houseleek (Fig. 65) and other *fleshy plants*. Here the close ranks of the thickened leaves are wholly above ground. Sometimes the deposit is all in the lower end of the leaf, and on the ground, or underneath, as in common *Bulbs*. Take a White Lily of the gardens, for example, in the fall, or in spring before it sends up the stalk of the season (Fig. 66). From the bottom of the bulb, roots descend into the soil to absorb moisture and other matters

65
Houseleek.

from it, while, above, it sends up leaves to digest and convert these matters into real nourishment. As fast as it is made, this nourishment is carried down to the bot-

3

tom of each leaf, which is enlarged or thickened for containing it. These thick
leaf-bases, or scales, crowded together, make up the bulb; all but its very short stem,
concealed within, which bears these scales above, and sends down the roots from
underneath. Fig. 67 shows one of the leaves of the season, taken
off, with its base cut across, that the thickness may be seen. After
having done its work, the blade dies off, leaving the thick base as
a bulb-scale. Every year one or more buds in the centre of the
bulb grow, feeding on the food laid up in the scales, and making
the stalk of the season, which bears the flowers, as in Fig. 1, 2.

78. An Onion is like a Lily-bulb, only each scale or leaf-base
is so wide that it enwraps all within, making coat after coat.

Bulb and lower Leaves of a Lily.   Leaf, lower end cut off.

79. In shrubs and trees a great quantity of nourishment, made the summer
before, is stored up in the young wood and bark of the shoots, the trunk, and the
roots. Upon this the buds feed the next spring; and this enables them to develop
vigorously, and clothe the naked branches with foliage in a few days; or with blossoms immediately following, as in the Horsechestnut; or with blossoms and foliage
together, as in Sugar Maple; or with blossoms before the leaves appear, as in Red
Maples and Elms. The rich mucilage of the bark of Slippery Elm, and the sweet
spring sap of Maple-trees, belong to this store, deposited in the wood the previous
summer, and in spring dissolved and rapidly drawn into the buds, to supply the early
and sudden leafing and blossoming.

80. In considering plants, as to "how they grow," it should be noticed that all of
them, from the Lily of the field to the tree of the forest, teach the same lesson of
industry and provident preparation. No great result is attained without effort, and

long preceding labor.   Not only was the tender verdure which, after a few spring showers and sunny days, is so suddenly spread out over field and forest, all prepared beforehand, — most of the leaves, even, made the summer before, and snugly packed away in winter-buds, — but the nourishment which enables them to unfold and grow so fast was also prepared for this purpose by the foliage of the year before, and laid up until it was wanted.   The grain grows with vigor, because fed with the richest products of the mother plant, the results of a former year's vegetation.   The Lily-blossom develops in all its glory without toil of its own, because all its materials were gathered from the earth and the air long before, by the roots and the leaves, manufactured by the latter into vegetable matter, and this stored up for a year or two under ground in the bottoms of the leaves (as starch, jelly, sugar, &c.), and in many cases actually made into blossoms in the dark earth, where the flower-buds lie slumbering in the protecting bulb through the cold winter, and in summer promptly unfold in beauty for our delight.

Analysis of the Section.

51. The seedling is a complete plant on the simplest scale ; in growth it merely increases its parts, and multiplies them in number, as fast as it makes materials for growth. 52. Simple stems, how formed and carried up, piece by piece. 53. Branches : 54. of Roots, how they differ from those, 55. of Stems. Where these arise from ; in what form they appear. 56. Buds, what they are. 57. Terminal Bud, what it makes. 58. Axillary Buds ; why so named ; what they make. 59. How branches are arranged, and what their arrangement depends upon: alternate; opposite. 60. The spray and buds of shrubs and trees in winter; Leaf-scars. 61. Why branches are not as regular and as many as the buds or leaves.

62. 63. The Duration and Character of Plants as affecting the way they grow. 64. Herbs. 65. Shrubs. 66. Trees. 67. Herbs are annuals, biennials, or perennials.

68. Annuals ; their mode of life ; character of their roots, intended only for absorbing; duration, &c.

69. Biennials ; how defined ; examples. 70. Character of their roots, and illustrations of their mode of life ; the first year, food made and stored up ; the second year, food expended, for what purpose. 71. How biennials may sometimes be made perennial, and annuals biennial. 72. The store of food may be kept in the leaves, or in the stems above ground ; Cabbage, &c.

73. Perennials ; what they are ; mode of life of perennial herbs from year to year ; accumulation of food in roots. 74. Accumulation of food in under-ground branches ; Tubers, as of Ground Artichoke. 75. Potato illustrated. 76. Accumulation in whole stems or branches under ground ; Rootstocks. 77. Accumulation of food in leaves, above ground, as in Houseleek ; or in the bottoms of leaves, usually under ground ; Bulbs ; as of Lily, and, 78. of Onion.

79. Food, how stored up in shrubs and trees, and for what purpose ; used in leafing and blossoming in spring. 80. A lesson taught by vegetation.

SECTION IV. — **Different Forms or Kinds of Roots, Stems, and Leaves.**

81. The Organs of Vegetation, or those that have to do with the life and growth of a plant, are only three, Root, Stem, and Leaf. And the plan upon which plants are made is simple enough. So simple and so few are the kinds of parts that one would hardly expect plants to exhibit the almost endless and ever-pleasing diversity they do. This diversity is owing to the wonderful variety of forms under which, without losing their proper nature, each of these three organs may appear.

82. The study of the different shapes and appearances which the same organ takes in different plants, or in different parts of the same plant, comparing them with one another, is called *Morphology*, and is one of the most interesting parts of Botany. But in this book for young beginners, we have only room to notice the commonest forms, and those very briefly, — although sufficiently to enable students to study all common plants and understand botanical descriptions. Those who would learn more of the structure and morphology of plants should study the Lessons in Botany.

§ 1.   *Of Roots.*

83. The **Root** is the simplest and least diversified of the three organs. Yet it exhibits some striking variations.

84. As to origin, there is the *primary* or original root, formed from the embryo as it grows from the seed, and the branches it makes. Annuals, biennials, and many trees are apt to have only such roots. But when any portion of their stems is covered by the soil, it makes *secondary* roots. These are roots which spring from the sides of the stem. Every one knows that most stems may be made to strike root when so covered and having the darkness and moisture which are generally needful for roots. Perennial herbs and most shrubs strike root naturally in this way under ground. All the roots of plants raised from tubers, rootstocks, and the like (74 – 76), are of this sort, and also of plants raised from slips or cuttings. In warm and damp climates there are likewise many

85. **Aerial Roots**, namely, roots which strike from the stem in the open air. In summer we often find them springing from the joints of the stalks of Indian Corn, several inches above the soil. Some of these reach the ground, and help to feed the plant. In the famous Banyan-tree of India aerial roots on a larger scale strike from the spreading branches, high up in the air, grow down to the ground and into it,

and so make props or additional trunks.   Growing in this way, there is no limit to
the extent of the branches, and a single Banyan will spread over several acres of
ground and have hundreds of trunks all made from aerial roots.

86. **Aerial Rootlets**, or such roots on a small scale, are produced by several woody
vines to climb by.   English Ivy, our Poison Ivy, and Trumpet-Creeper are well-
known cases of the sort.

87. **Air-Plants.**   Roots which never reach the ground are also produced by certain
plants whose seeds, lodged upon the boughs or trunks of trees, high up in the
air, grow there, and make an
Epiphyte, as it is called (from
two Greek words meaning
a plant on a plant), or an
Air-Plant.   The latter name
refers to the plant's getting
its living altogether from the
air ; as it must, for it has no
connection with the ground
at any time.   And if these
plants can live on air, in this
way, it is easy to understand
that common vegetables get
part of what they live on directly from the air.   In warm
countries there are many very
handsome and curious air-
plants of the Orchis family.
A great number are culti-
vated in hot-houses, merely
fixed upon pieces of wood
and hung up.   They take no

Air-plants of the Orchis family.

nourishment from the boughs of the tree they happen to grow upon.

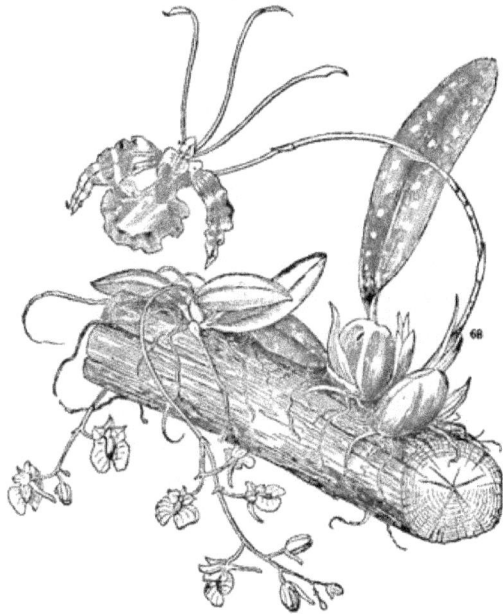

88. **Parasitic Plants** are those which strike their roots, or what answer to roots,
into the bark or wood of the species they grow on, and feed upon its sap.   The
Mistletoe is a woody parasitic plant, which engrafts itself when it springs from the
seed upon the branches of Oaks, Hickories, or other trees.   The Dodder is a com-

mon parasitic herb, consisting of orange-color or whitish stems, looking like threads of yarn. These coil round the stalks of other plants, fasten themselves by little suckers in place of roots, and feed upon their juices. Living as such a plant does by robbing other plants of their prepared food, it has no leaves of its own, except little scales in their place, and has no need of any.

89. **Shapes and Uses of Roots.** Common roots, however, grow in the soil. And their use is to absorb moisture and other matters from the soil, and sometimes to hold prepared food until it is wanted for use, as was explained in the last section (70, 73). Those for absorbing are

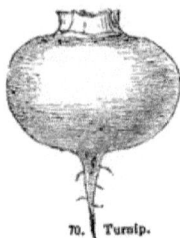

Fibrous roots, namely, slender and thread-shaped, as in Fig. 48, 56, and generally branching. Very slender roots of the sort, or their branches, are called *Rootlets;* and these do most of the absorbing. The roots of annuals are mostly fibrous, as they have nothing to do but to absorb; and so are the smaller branches of the roots of shrubs, trees, and other plants.

70. Turnip.

Fleshy roots are those of herbs which form a thick and stout body, from having much nourishment deposited in them. They belong particularly to biennial herbs (69), and to many perennials (73). Some sorts have names according to their shapes. The root is a

Tap-root, when of one main body, and tapering downwards to a point; as that of a Carrot (Fig. 71), and of a seedling Oak (Fig. 41). And a tap-root is

Conical, when stout, and tapering gradually from the upper end to a point below; as a carrot (Fig. 71), parsnip, or beet.

Spindle-shaped, when thicker in the middle, and tapering upwards as well as downwards, like a radish (Fig. 57); and

Turnip-shaped, or *Napiform,* when wider than long, or with a suddenly tapering tip, as a turnip (Fig. 70). Roots are

71. Carrot.

Clustered or *Fascicled* when, instead of one main root, there are several or many of about the same size; as in Indian Corn (Fig. 48), and other grain (Fig. 56). Here the clustered roots are *fibrous,* being for absorbing only. When such roots, or some of them, are thick and fleshy, as they are when used as storehouses of food, they become *Tuberous.* The roots of the Dahlia, for instance (Fig. 58), are *clustered* and *tuberous,* or *tuber-like.*

§ 2. *Of Stems.*

90. **Forms or Kinds of Stems.** Differences in the size and consistence of stems, such as distinguish plants into *herbs*, *shrubs*, and *trees*, have already been noticed, in paragraphs 64, 65, and 66. A stem is

Herbaceous, when it belongs to an herb, that is, has very little wood in its composition, and does not live over winter above ground:

Shrubby, when it belongs to a shrub, or is woody:

Arboreous or *Arborescent*, when the plant is a tree, or like a tree; that is, when it is tall and grows by a single *trunk.*

91. The peculiar straw-stem of a grass or grain is named a *Culm*. It is generally hollow, except at the joints, which are hard and solid; but in Indian Corn, Sugar-Cane, and some other Grasses, it is not at all hollow.

92. As to the mode of growth or the direction it takes in growing, the stem is

Erect or *Upright*, when it grows directly upwards, or nearly so:

Ascending, when it rises upwards at first in a slanting direction:

Declined or *Reclined*, when turned or bent over to one side:

Decumbent, when the lower part reclines on the ground, as if too weak to stand, but the end turns upwards more or less:

Procumbent or *Trailing*, when the whole stem trails along the ground:

Prostrate, when it naturally lies flat on the ground:

Creeping or *Running*, when a trailing or prostrate stem strikes root along its lower side, where it rests on the ground:

Climbing, where it rises by laying hold of other objects for support; either by tendrils, as in the Pea, Gourd, and Grape-Vine; or by twisting its leafstalks around the supporting body, as in the Virgin's Bower; or by rootlets acting as holdfasts, as in the Ivy and Trumpet-Creeper (86):

Twining, when stems rise by coiling themselves spirally around any support, as in the Morning-Glory (Fig. 4), Hop, and Bean.

93. Several sorts of branches are different enough from the common to have particular names. Indeed, some are so different, that they would not be taken for branches without considerable study. Such, for instance, as

94. **Thorns or Spines.** Most of these are imperfect, leafless, hardened, stunted branches, tapering to a point. That they are branches is evident in the Hawthorn and similar trees, from their arising from the axil of leaves, as branches do. And on Pear-trees and Plum-trees many shoots may be found which begin as a leafy

branch, but taper off into a thorn. *Prickles*, such as those on the stems of Roses and Brambles, must not be confounded with thorns. These are growths from the bark (like hairs or bristles, only stouter), and peel off with it; while thorns are connected with the wood.

95. **Tendrils**, such as those of the Grape-Vine, Virginia Creeper (Fig. 72), and the Melon or Squash, are very slender, leafless branches, used to enable certain plants to climb. They grow out straight or nearly so until they reach some neighboring support, such as a stem, when the end hooks around it to secure a hold, and the whole tendril then shortens itself by coiling up spirally, so drawing the growing shoot nearer to the

Tendrils of Virginia Creeper.

72   73

supporting object. When the Virginia Creeper climbs the side of a building, the face of a rock, or the smooth bark of a tree, which the tendrils cannot lay hold of in the usual way, their tips expand into a flat plate (as shown in Fig. 73, the ends of a tendril magnified), which adheres very firmly to the surface. This enables the plant to climb up a smooth surface by tendrils, just as the Ivy and Trumpet-Creeper climb by rootlets (86).

96. **Peduncles** or **Flower-stalks** are a kind of branches, or stems, as is clear from their situation. They are either a continuation of the stem, as in the Lily of the Valley and the Chalcedonian Lily, represented on the first page; or else they rise out of the axil of a leaf, as in the Morning-Glory (Fig. 4). Plainly, whatever comes from the axil of a leaf must be of the nature of a branch. So

97. **Buds**, that is axillary buds, are undeveloped branches, as already explained in paragraphs 55 to 58.

98. The following kinds of branches are all connected with the ground in some way, and most of them act in such a way as to make new plants.

99. **A Stolon** is a branch which reclines on the ground, or bends over to it, and strikes root (Fig. 74). Currant-bushes spread naturally by stolons, and so does White Clover. The gardener imitates the process where it does not naturally occur, or facilitates it where it does, by bending branches to the ground, and pinning them down, when they strike root where they are covered by the soil, and then the branch, having leaves and roots of its own, may be separated as an independent plant. In this way the gardener multiplies many plants by *layering* which he cannot so readily propagate by seed.

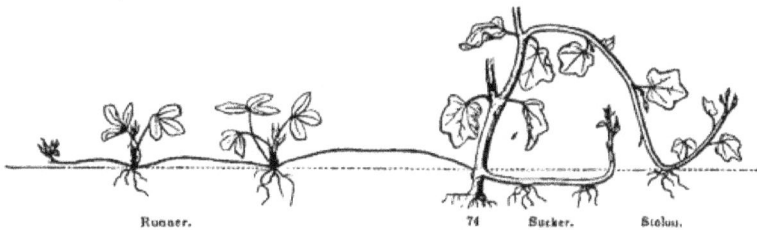

Runner. 74 Sucker. Stolon.

100. **A Runner** (Fig. 74) is a very slender, thread-like, leafless stolon, much like a tendril, lying on the ground, and rooting and budding at the point; so giving rise to a new plant at some distance from the parent, and connected with it during the first year. But the runner dies in winter and leaves the young plant independent. The Strawberry-plant affords the most familiar illustration of runners. Each plant or offshoot, as soon as established, sends out runners of its own, which make new plants at their tip. In this way a single Strawberry-plant produces a numerous progeny in the course of the summer, and establishes them at convenient distances all around.

101. **A Sucker** (Fig. 74) is a branch which springs from a parent stem under ground, where it makes roots of its own, while farther on it rises above ground into a leafy stem, and becomes an independent plant whenever the connection with the parent stem dies or is cut off. It is by suckers that Rose and Raspberry bushes multiply and spread so "by the root," as is generally said. But that these subterranean shoots are stems, and not roots (though they produce roots), will plainly appear by uncovering them.

102. **An Offset** is a short branch, next the ground or below its surface, like a short stolon or sucker, bearing a tuft of leaves at the end, and taking root where this

rests on the soil; as in the Houseleek (Fig. 65), where one plant will soon produce a cluster of young plants or offsets all around it.

103. **A Rootstock** is any kind of horizontal stem or branch growing under ground. Slender rootstocks occur in the subterranean part of the suckers of Roses, of Peppermint, or of Canada Thistle, and of Quick-Grass or Couch-Grass (Fig. 75), which spreads so widely, and becomes so troublesome to farmers. They are well distinguished from roots by the leaves which they bear at every joint, in the form of scales, and by the buds which they produce, one in the axil of each scale. These buds, which are very tenacious of life, are what renders the plant so exceedingly difficult to destroy. For ploughing and hoeing only cut up the rootstock into pieces, each with a tuft of roots ready formed and with a bud to each joint, all the more ready to grow for the division. So that the attempt to destroy Quick-Grass by cutting it up by the roots (as these shoots are called), unless the pieces are carefully

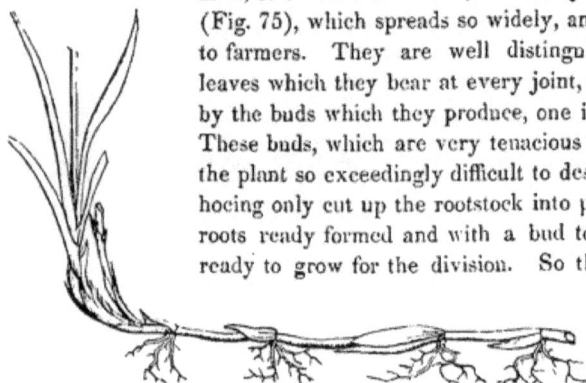

75
Rootstock of Quick-grass.

taken out of the soil, is apt to produce many active plants in place of one.

104. Thickened or fleshy rootstocks, such as those of Solomon's Seal (Fig. 63) and Iris (Fig. 64), have already been illustrated (76).

105. **A Tuber** is a rootstock thickened at the end, as already explained in the Potato and Ground Artichoke (74, 75, Fig. 59, 60). The eyes of a tuber are lively buds, well supplied with nourishment for their growth.

106. **A Corm** or **Solid Bulb**, as of Gladiolus and Crocus (Fig. 76), is a sort of rounded tuber. If well covered with thick scales it would become

107. **A Bulb.** This is a (mostly subterranean) stem, so short as to be only a flat plate, producing roots from its lower surface and above covered with thickened scales, — as was fully explained in the last section (77).

76
Corm of Crocus, with buds.

108. Bulbs are *scaly*, as in the Lily (Fig. 66), when the scales are narrow; or *coated*, as an onion, when the scales enwrap each other, and form coats.

109. **Bulblets** are little bulbs, or fleshy buds, formed in the axils of leaves above ground, as in the Bulb-bearing Lily. Or in some Leeks and Onions they take the place of flower-buds. Falling off, they take root and grow into new plants.

110. **The Internal Structure of Stems.** Plants are composed of two kinds of material, namely, *Cellular Tissue* and *Wood*. The former makes the softer, fleshy, and pithy parts; the latter forms the harder, fibrous, or woody parts. The stems of herbs contain little wood, and much cellular tissue; those of shrubs and trees abound in the woody part.

111. There are two great classes of stems, which differ in the way the woody part is arranged in the cellular tissue. They are named the *Exógenous*, and the *Endógenous*.

112. For examples of the first class we may take a Bean-stalk, a stem of Flax, Sunflower, or the like, among herbs, and for woody stems any common stick of wood. For examples of the second class take an Asparagus-shoot or a Cornstalk, and in trees a Palm-stem. These names express the different ways in which the two kinds grow in thickness when they live more than one year. But the difference between the two is almost as apparent the first year, and in the stems of herbs, which last only one year.

113. **The Endogenous Stem.** *Endógenous* means "inside-growing." Fig. 77 shows an Endogenous stem in a Cornstalk, both in a cross-section, at the top, and also split down lengthwise. The peculiarity is that the wood is all in separate threads or bundles of fibres running lengthwise, and scattered among the cellular tissue throughout the whole thickness of the stem. On the cross-section their cut ends appear as so many dots; in the slice lengthwise they show themselves to be threads or fibres of wood. Fig. 78 is a similar view of a Palm-stem (namely, of our Carolina Palmetto, of which whole trees are represented in Fig. 79). It shows the endogenous plan in a stem several years old. Here the bundles of wood are merely increased very much in number, new threads having been formed throughout intermixed with the old, and any in-crease in diameter that has taken place is from a general distention or enlargement

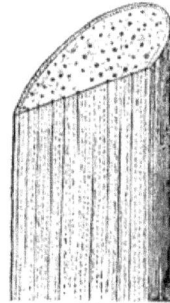

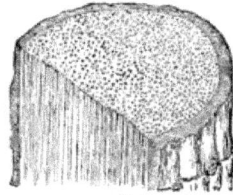

Endogenous Stems.

of the whole. Such stems may well enough be called *inside-growers*, because their wood increases in amount, as they grow older, by the formation of new threads or fibres of wood within or among the old.

114. Moreover, endogenous stems are apt to make few or no branches. Asparagus is the only common example to the contrary ; that branches freely. But the stalks of Corn and other grain, and those of Lilies (Fig. 1, 2) and the like, seldom branch until they come to flower ; and Palms are trees of this sort, with perfectly simple or branchless trunks, rising like columns, and crowned with a tuft of conspicuous and peculiar foliage, which all comes from the continued growth of a terminal bud.

115. The Exogenous Stem is the kind we are familiar with in ordinary wood. But it may be observed in the greater part of our herbs as well. It differs from the other class, even at the beginning, by the wood all occupying a certain part of the stem, and by its woody bundles soon appearing to run together into a solid layer. This layer of wood,

Palmettos of various ages, and a Yucca, y.

whether much or little, is always situated around a central part, or *pith*, which has no wood in it, being pure cellular tissue, and is itself surrounded by a bark which is mainly or at first entirely cellular tissue. So that a slice across an exogenous stem always has a separate cellular part, as bark, on the circumference, then a ring of wood, and in the centre a pith ; as is seen in Fig. 80, representing a piece

of Flax-stem magnified; and also in Fig. 81, which shows the same structure in a woody stem, namely, in a shoot of Maple of a year old, cut both crosswise and lengthwise.

116. The difference becomes still more marked in stems more than one year old. During the second year a new layer of wood is formed outside of the first one, between it and the bark; the third year, another layer outside of the second, and so on, a new layer being formed each year outside of that of the year before. The increase is all on the surface, and buries the older wood deeper and deeper in the trunk. For this reason such stems are said to be *exógenous* or outside-growing (from two Greek words which mean just this), a new layer being added to the wood on the outside each year as long as the tree or shrub lives. And so the oldest wood, or *Heart-wood*, is always in the centre, and the newest and freshest, the *Sap-wood*, at the circumference, just beneath the bark.

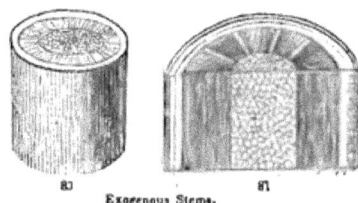

Exogenous Stems.

117. The heart-wood is dead, or soon becomes so. The sap-wood is the only active part; and this, with the inner bark, which is renewed from its inner face every year, is all of the trunk that is concerned in the life and growth of the tree.

118. Plants with exogenous or outside-growing stems, especially those that live year after year, almost always branch freely. All common shrubs and trees of the exogenous class make a new set of branches every year, and so present an appearance very different from that of most of those of the endogenous or inside-growing class.

§ 3.   *Of Leaves.*

119. Leaves exhibit an almost endless variety of forms in different plants; and their forms afford easy marks for distinguishing one species from another. So the different shapes of leaves are classified and named very particularly, — which is a great convenience in describing plants, as it enables a botanist to give a correct idea of almost any leaf in one or two words. We proceed to notice some of the principal kinds.

120. Their Parts. A leaf with all its parts complete has a *Blade*, a *Footstalk*, and a pair of *Stipules* at the base of the footstalk. Fig. 82 shows all three parts

in a Quince-leaf: *b*, the blade ; *p*, the footstalk ; and *st*, the stipules, looking like a pair of little blades, one on each side of the stalk.   But many leaves have no stipules ; many have no footstalk, and then the blade sits directly on the stem (or is *sessile*), as in Fig. 138.   Some leaves even have no blade; but this is uncommon; for in foliage the blade is the essential part.   There-fore, in describing the shape of leaves, it is always the blade that is meant, unless something is said to the contrary.

121. Leaves are either *simple* or *compound*.   They are *simple* when the blade is all of one piece ; *com-pound*, when of more than one piece or blade.   Fig. 128 to 132, and 133, are examples of compound leaves, the latter very compound, having as many as eighty-one little blades.

122. **Their Structure and Veining.**   Leaves are com-posed of the same two kinds of material as stems (110), namely, of wood or fibre, and of cellular tissue.   The woody or fibrous part makes a framework of ribs and veins, which gives the leaf more strength and toughness than it would otherwise have.   The cellu-lar tissue forms the *green pulp* of the leaf.   This is spread, as it were, over the framework, both above and below, and supported by it ; and the whole is protected by a transparent skin, which is termed the *Epidermis.*

123. **Ribs.**   The stouter pieces or timbers of the framework are called *Ribs*. In the leaf of the Quince (Fig. 82), Pear, Oak (Fig. 120), &c. there is only a single main rib, running directly through the middle of the blade from base to point ; this is called the *Midrib*.   But in the Mallow, the Linden (Fig. 83), the Maple (Fig. 84), and many others, there are three, or five, or seven ribs of nearly the same size. The branches of the ribs and the branchlets from them are called

124. **Veins and Veinlets.**   The former is the general name for them ; but the finest branches are particularly called *Veinlets*.   Straight and parallel veins or fine ribs, like those of Indian Corn, or of any Grass-leaf, or of the Lily of the Valley (Fig. 3, 85), are called *Nerves*.   This is not a sensible name, for even if in some degree like the nerves of animals in shape, they are not in the least like them in use.

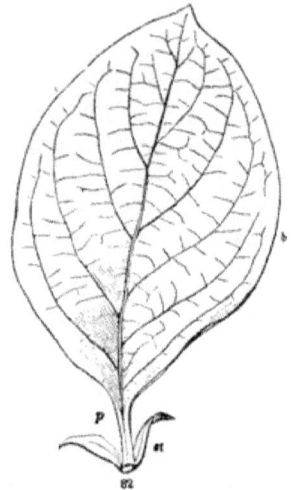

Nor are what we call *veins* to be likened particularly to the bloodvessels of animals. But this name is not so bad; for the minute fibres which, united in bundles, make up the ribs and veins, are hollow tubes, and serve more or less for conveying the sap.

125. As to the *veining*, or the arrangement of the framework in the blade, leaves are divided into two classes, viz.. 1st, the *Netted-veined* or *Reticulated*, and, 2d, the *Parallel-veined* or *Nerved*.

126. **Netted-Veined** or **Reticulated** leaves are those in which the veins branch off from the rib or ribs, and divide again and again, and some of the veins and veinlets run into one another, so forming reticulations or meshes of network throughout the leaf. This is shown in the Quince-leaf (Fig. 82); also in the Linden or Basswood (Fig. 83), and the Maple (Fig. 84), where the finer meshes appear in one or two of the leaves.

127. Netted-veined leaves belong to plants which have a pair of seed-leaves to their embryo (48), and stems

83. Linden.   Netted-veined Leaves of   84. Maple.

of the exogenous structure (115). That is, these three kinds of structure, in embryo, stem, and leaf, generally go together.

128. **Parallel-veined** or **Nerved** leaves are those in which the ribs and veins run side by side without branching (or with minute cross-veinlets, if any) from the base to the point of the blade, as in Indian-Corn, Lily of the Valley (Fig. 85), &c., or sometimes from the midrib to the margins, as in the Banana and Calla (Fig. 86). Such parallel veins have been called *Nerves*, as just explained (124). Leaves of this sort belong to plants with one cotyledon to their embryo (47), and with endogenous stems (113).

129. *Parallel-veined* leaves, we see, are of two sorts; — 1. those with the veins or nerves all running from the base of the leaf to the point (Fig. 85) ; and, 2. those where they mostly run from the midrib to the margin, as in Fig. 86. *Netted-veined* leaves likewise are of two sorts, the *Feather-veined* and the *Radiate-veined.*

130. *Feather-veined* (also called *pinnately veined*) leaves are those in which the main veins all spring from the two sides of one rib, viz. the midrib, like the plume of a feather from each side of the shaft. Figures 82, 88 – 97, 120, 122, &c. represent feather-veined leaves.

131. *Radiate-Veined* (also called *palmately veined*) leaves are those which have three or more main ribs rising at once from the place where the footstalk joins the blade, and commonly diverg-

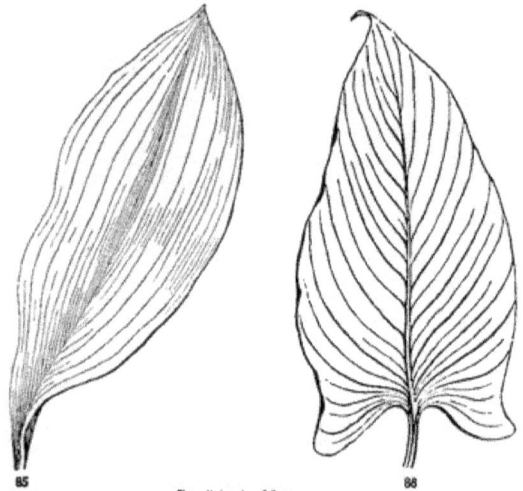

Parallel-veined Leaves.

ing, like rays from a centre ; the veins branching off from these. Of this sort are the leaves of the Maple (Fig. 84), Mallow, Currant, Grape-Vine, and less distinctly of the Linden (Fig. 83). Such leaves are generally roundish in shape. It is evident that this kind of veining is adapted to round leaves, and the other kind for those longer than wide.

132. **Shapes of Leaves.** As to general shape, the following are the names of the principal sorts. (It will be a good exercise for students to look up examples which fit the definitions.)

Linear ; narrow, several times longer than wide, and of about the same width throughout, as in Fig. 87.

Lance-shaped or *Lanceolate ;* narrow, much longer than wide, and tapering upwards, or both upwards and downwards, as in Fig. 88.

Oblong ; two or three times longer than broad, as in Fig. 89.

Oval ; broader than oblong, and with a flowing outline, as in Fig. 90.

Ovate ; oval, but broader towards the lower end; of the shape of a hen's egg cut through lengthwise, as in Fig. 91.

Orbicular or *Round ;* circular or nearly circular in outline, as in Fig. 93.

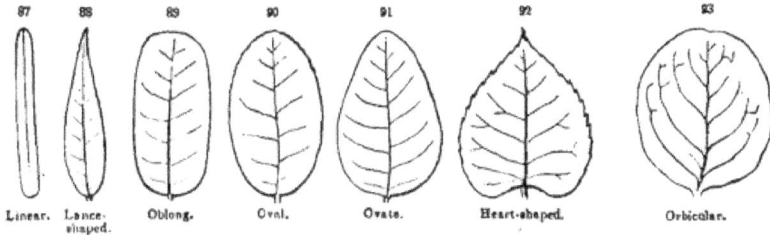

Linear. Lance-shaped. Oblong. Oval. Ovate. Heart-shaped. Orbicular.

133. Some leaves taper downwards more than upwards. Of these the commonest forms are the

Oblanceolate, or *Inversely lance-shaped ;* that is, shaped like a lance with the point downwards, as in Fig. 94.

Spatulate ; roundish above, and tapering into a long and narrow base, like the old form of the apothecary's spatula, Fig. 95.

Obovate, or *Inversely ovate* ; that is, ovate with the narrow end at the bottom of the leaf, as in Fig. 96.

Cuneate or *Wedge-shaped ;* like the last, but with the sides narrowing straight down to the lower end, in the shape of a wedge, as in Fig. 97.

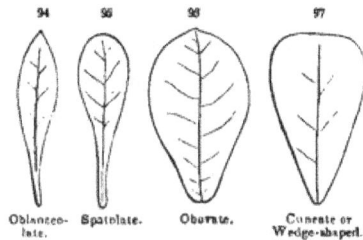

Oblanceo- Spatulate. Obovate. Cuneate or
late. Wedge-shaped.

134. Of course these shapes all run into one another by imperceptible degrees in different cases. The botanist merely gives names to the principal grades. Intermediate shapes are described by combining the names of the two shapes the leaf in question most resembles. For example : —

Lance-linear, or *linear-lanceolate,* means between linear and lance-shaped.

Lance-oblong, or *oblong-lanceolate,* means between oblong and lanceolate in shape.

Ovate-lanceolate, between ovate and lance-shaped ; and so on.

135. Or else a qualifying word may be used, as *somewhat* ovate, *slightly* heart-shaped, and the like. Thus, Fig. 92 is ovate in general form, but with the base a little notched, i. e. *somewhat* heart-shaped. It is one of the kinds which depend upon

4

136. *The shape at the base.* This is concerned in all the following sorts: —

Heart-shaped, or *Cordate;* when of the shape in which a heart is painted, the base having a recess or notch, as in Fig. 98.

Kidney-shaped, or *Reniform;* like heart-shaped, but rounder, and broader than long, as in Fig. 99.

Auricled, or *Eared;* having a small projection or *lobe* on each side at the base, like a pair of ears, as in Fig. 101.

Arrow-shaped, or *Arrow-headed;* when such lobes at the base are

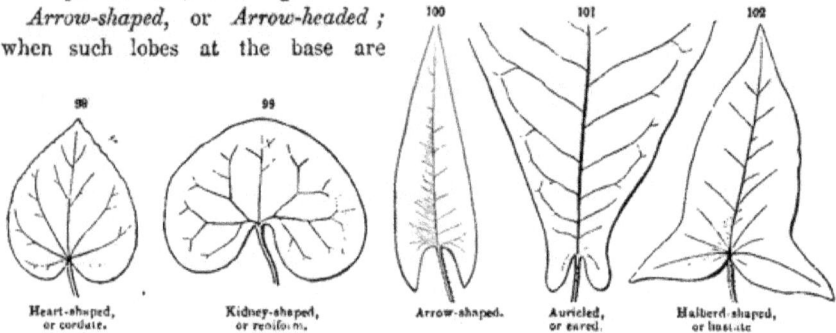

Heart-shaped,   Kidney-shaped,   Arrow-shaped.   Auricled,   Halberd-shaped,
or cordate.   or reniform.   or eared.   or hastate

pointed and turned backwards, like the base of an arrow-head, as in Fig. 100.

Halberd-shaped, or *Hastate;* when such lobes point outwards, giving the whole blade the shape of the halberd of the olden time, as in Fig. 102.

Shield-shaped, or *Peltate;* when the footstalk is attached to some part of the lower face of the blade, which may be likened to a shield borne by the hand with the arm extended. Fig. 104 represents the shield-shaped leaf of a Water-Penny-wort. Fig. 103 is the leaf of another species, which is not shield-shaped. A comparison of the two shows how the shield-shaped leaf is made.

137. *As to the Apex or Point,* we have the following terms, the first six of which apply to the base as well as to the apex of a leaf: —

Pointed, *Taper-pointed,* or *Acuminate;* narrowed into a tapering tip, as in Fig. 105.

Acute; ending in an acute angle, Fig. 106.

Round-kidney-shaped.   Shield-shaped.

Obtuse; ending in an obtuse angle, or with a blunt or rounded apex; as in Fig. 107.

Truncate; as if cut off square at the apex, as in Fig. 108.

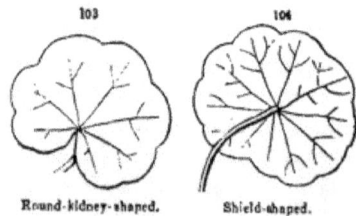

Retuse ; having a blunt or rounded apex slightly indented, as in Fig. 109.

Emarginate, or *Notched ;* as if a notch were cut out of the apex ; Fig. 110.

Obcordate, or *Inversely heart-shaped ;* that is, with the strong notch at the apex instead of the base, as in Fig. 111 and the leaflets of White Clover.

Cuspidate ; tipped with a rigid or sharp and narrow point, as in Fig. 112.

Mucronate ; abruptly tipped with a short and weak point, like a small projection of the midrib, as in Fig. 113.

Awned, Awn-pointed, or *Aristate ;* tipped with a long bristle-shaped appendage, like the beard (*awn*) of Oats, &c.

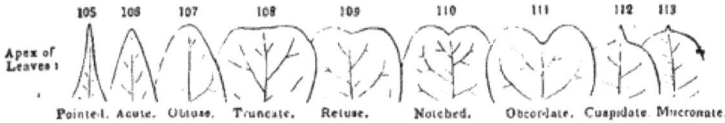

Apex of Leaves :

Pointed. Acute. Obtuse. Truncate. Retuse. Notched. Obcordate. Cuspidate. Mucronate.

138. *As to the margin*, whether whole, toothed, or cut, leaves are said to be

Entire ; when the margin is an even line, as in Fig. 99 to 102.

Toothed ; when beset with teeth or small indentations ; of this there are two or three varieties, as,

Serrate or *Saw-toothed ;* when the teeth turn forwards, like those of a saw, as in Fig. 114.

Dentate ; when they point outward, as in Fig. 115.

Crenate ; when *scalloped* into broad and rounded teeth, as in Fig. 116.

Wavy (*Repand* or *Undulate*); when the margin bends slightly in and out, as in Fig. 117.

Sinuate ; strongly wavy or sinuous, as in Fig. 118.

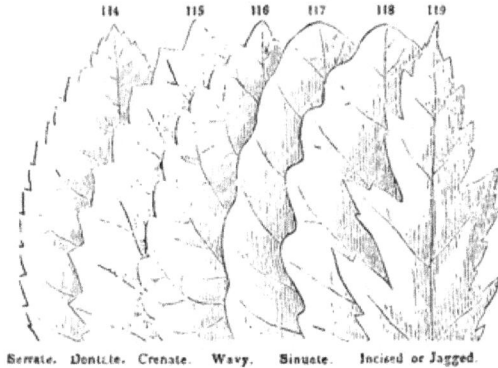

Serrate. Dentate. Crenate. Wavy. Sinuate. Incised or Jagged.

Incised or *Jagged ;* cut into deep and irregular, jagged teeth or incisions, as in Fig. 119. This leads to truly

139. **Lobed** or **Cleft Leaves**, &c.. those with the blade cut up, as it were, into parts, i. e. *lobes* or *divisions.* In a general way, such leaves are said to be *lobed ;* and the

number of projecting parts, or lobes, may be expressed by saying *two-lobed, three-lobed* (Fig. 121), &c., according to their number. Or, more particularly, a leaf is

Lobed ; when the pieces are roundish, or the incisions open or blunt, as in Fig. 120, 121 ; and

Cleft ; when cut about half-way down, with sharp and narrow incisions, as in Fig. 122, 123 ; and so *two-cleft, three-cleft, five-cleft,* &c., according to the number.

Parted ; when the cutting extends almost through, as in Fig. 124, 125. And we say *two-parted, three-parted,* &c., to express the number of the parts.

Divided ; when the divisions go through to the base of the leaf (as in Fig. 127), or to the midrib (as in Fig. 126), which cuts up the blade into separate pieces, or nearly so.

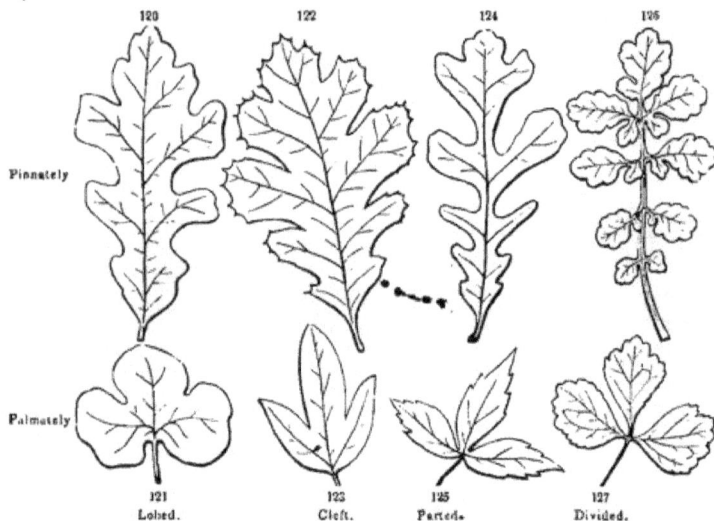

140. As the cutting is always *between* the veins or ribs, and not across them, the arrangement of the lobes depends upon the kind of veining. Feather-veined leaves have the incisions all running in towards the midrib (as in the upper row of figures), because the principal veins all spring from the midrib ; while radiate or palmately veined leaves have them all running towards the base of the blade, where the ribs all spring from the footstalk, as in the lower row of figures. So those of

the upper row are called *pinnately lobed, cleft, parted,* or *divided,* as the case may be, and those of the lower row *palmately lobed, cleft,* &c.   The number of the lobes or pieces may also be expressed in the same phrase.   Thus, Hepatica has a *palmately three-lobed* leaf (Fig. 121) ; the Red Maple a *palmately five-cleft* leaf (Fig. 84), and so on.

141. In this way almost everything about the shape and veining of a leaf may be told in very few words.   How useful this is, will be seen when we come to study plants to find out their names by the descriptions.

142. All these terms apply as well to the lobes or parts of a leaf, when they are themselves toothed, or lobed, or cleft, &c.   And they also apply to the parts of the flower, and to any flat body like a leaf.   So that the language of Botany, which the student has to learn, does not require so very many technical words as is commonly supposed.

143. **Compound Leaves** (121) are those which have the blade cut up into two or more separate smaller blades.   The separate blades or pieces of a compound leaf are called *Leaflets.*   The leaflets are generally *jointed* with the main footstalk, just as that is jointed with the stem, and when the leaf dies the leaflets fall off separately.

144. There are two kinds of compound leaves, the *pinnate* and the *palmate.*

145.   *Pinnate* leaves have their leaflets arranged along the sides of the main footstalk, as in Fig. 128, 129, 130.

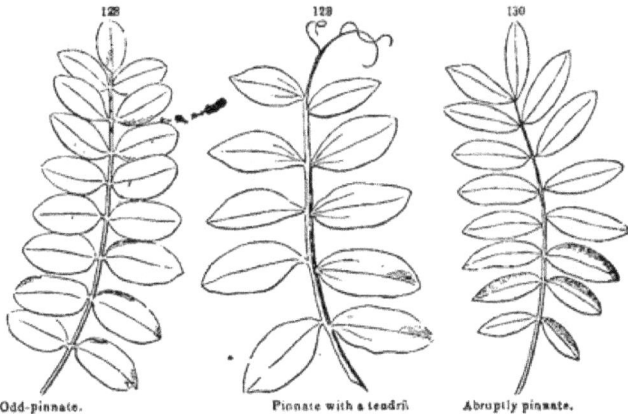

Odd-pinnate.   Pinnate with a tendril.   Abruptly pinnate.

146. *Palmate* (also called *Digitate*) leaves bear their leaflets all at the very end of the footstalk ; as in Fig. 131.

147. There are several varieties of pinnate leaves.   The principal sorts are : —

Interruptedly pinnate, when some of the leaflets of the same leaf are much smaller than the rest, and placed between them, as in the Water Avens.

131. Palmate leaf, of 5 leaflets.

Abruptly pinnate, when there is no odd leaflet at the end, as in Honey-Locust, Fig. 130.

Odd-pinnate, when there is an odd leaflet at the end, as in the Common Locust (Fig. 128) and in the Ash.

Pinnate with a tendril, when the footstalk is prolonged into a tendril, as in Fig. 129, and all of the Pea tribe.

148. Pinnate leaves may have many or few leaflets. The Bean has pinnate leaves of only three leaflets.

149. Palmate leaves generally have few leaflets; there is not room for many on the very end of the footstalk. Common Clover has a palmate leaf of three leaflets (Fig. 136); Virginia Creeper, one of five leaflets (Fig. 72), as well as the Buckeye (Fig. 131); while the Horsechestnut has seven, and some Lupines from eleven to seventeen.

150. **Twice or Thrice Compound Leaves** are not uncommon, both of the pinnate and of the palmate sorts. While some leaves of Honey-Locust are only once pinnate, as in Fig. 130, others are doubly or *twice pinnate,* as in Fig. 132. Those of many Acacias are *thrice pinnate.* Fig. 133 represents one of the root-leaves of Meadow-Rue, which is of the palmate kind, and its general footstalk is divided into threes for four times in succession, making in all eighty-one leaflets! When a leaf is divided three or four times, it is said to be *decompound.* This is *ternately* decompound, because it divides each time into threes.

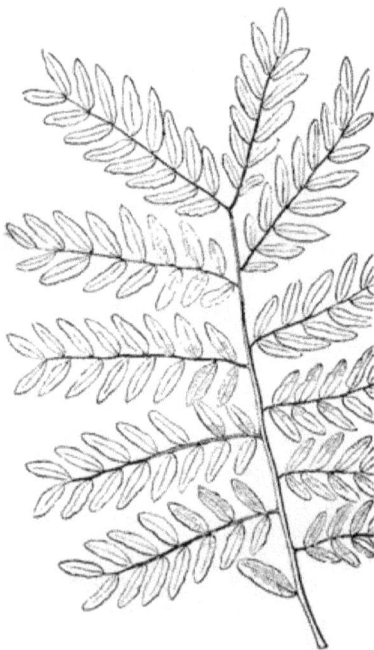

132. A twice-pinnate leaf of Honey-Locust.

151. Leaves without Distinction of Footstalk and Blade, or with no very obvious distinction of parts. Of this kind, among several others, may be mentioned, —

Needle - shaped leaves, such as those of Pine-trees and Larches (Fig. 134). These are long, slender, and rigid, and often with little if any distinction of sides.

Awl-shaped or *Subulate* leaves are those which from a broadish base taper into a sharp and rigid point, like

133
Ternately decompound, or four times compound leaf.

one sort of those of the Red Cedar and Arbor Vitæ (Fig. 135, those on the larger branchlets). Those on other branchlets, as at *a*, are shorter, blunt, and *scale-shaped.*

Thread-shaped or *Filiform* leaves ; round and stalk-like, as those of the Onion.

Equitant leaves, like those of Iris (Fig. 64), which are folded together lengthwise, as may be seen at the base, where they override each other. They grow upright, with their faces looking

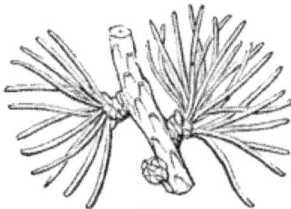

134
Needle-shaped leaves of Larch.

135
Awl-shaped leaves, &c.

horizontally, instead of having an upper and a lower surface, as most leaves do.

152. **Stipules,** as already explained (120), are a pair of appendages at the base of the leaf, one on each side. These often grow fast to the base of the leafstalk, as they do in the Rose and in Clover (Fig. 136; *st,* the stipules). Or they may join with each other and form a kind of sheath round the stem, as they do in the Buttonwood and in Polygonum (Fig. 137). Many leaves have no stipules at all. In many cases they fall off very early, especially those that serve for bud-scales, as in Magnolia.

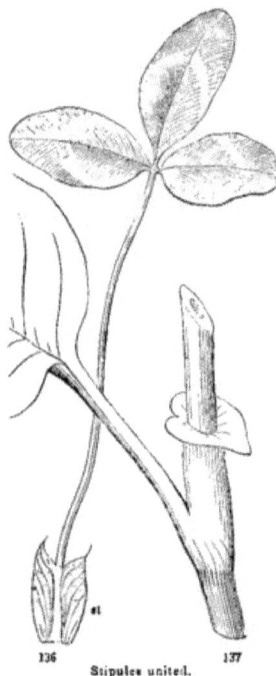

153. **The Arrangement of Leaves** on the stem has already been explained as to the two principal ways (59). Leaves are either

Alternate, when they follow each other one by one, as in the Morning-Glory (Fig. 4) and the Linden (Fig. 83); or

Opposite, when in pairs, that is, two on each joint of stem, one opposite the other, as in Maples (Fig. 84). To these may be added a third, but less common arrangement, viz. the

Whorled ; where there are three, four, or more leaves on the same joint of stem, forming

a circle or *whorl;* as in Madder and Bedstraw (Fig. 137'). But this is only a variety of the opposite mode.

136   137 Stipules united.   137' Whorled leaves.

Analysis of the Section.

81. Vegetation very simple in plan, very diversified in particulars. 82. The study of the forms of the organs is Morphology.

83 – 89. Roots, their forms and kinds. 84. Primary or original; secondary; how they originate. 85. Aerial roots. 86. Aerial rootlets. 87. Air-Plants ; how they live. 88. Parasitic Plants, their economy. 89. Shapes of roots: fibrous; fleshy; the principal sorts.

90. Forms or kinds of stem; herbaceous, shrubby, arboreous. 91. Culm or straw-stem. 92. Directions or positions of stems. 93. Peculiar sorts. 94. Thorns or Spines, how shown to be branches;

Prickles. 95. Tendrils. 96. Peduncles or Flower-stalks. 97. Buds. 98. Branches connected with the ground. 99. Stolons. 100. Runners. 101. Suckers. 102. Offsets. 103. Rootstocks. 104. Fleshy Rootstocks. 105. Tubers. 106. Corms. 107. Bulbs; 108. scaly and coated. 109. Bulblets.

110. Internal Structure of Stems; Cellular Tissue; Wood. 111. The two classes of stems. 112. Examples, both in herbs and trees. 113. Endogenous stem; how its wood is arranged. 114. External appearance and growth. 115. Exogenous stem; common wood. 116. How it increases in diameter year after year: Sap-wood and Heart-wood. 117. The latter dead, the former annually renewed. 118. External appearance and mode of growth.

119. Leaves ; their varieties, why useful to learn. 120. Their parts : Blade, Footstalk, Stipules. 121. Simple and Compound. 122. Structure and Veining of leaves: woody or fibrous part; cellular tissue or green pulp ; Epidermis or Skin. 123. Ribs. 124. Veins and Veinlets; Nerves, so called. 125. Two kinds of veining. 126. Netted-veined or Reticulated. 127. Class of plants that have this kind of veining. 128. Parallel-veined or Nerved ; class of plants that have this kind of veining. 129. Both kinds of two sorts. 130. Feather-veined or Pinnately veined. 131. Radiate-veined or Palmately veined.

132 Shapes of leaves enumerated; as to general outline. 133. Those that taper downward. 134, 135. Intermediate shapes, how expressed. 136. Shapes depending upon the base. 137. Forms of apex. 138. As to margin or toothing, &c. 139. Lobing or division. 140. How this is related to the veining; how both the kind of lobing and the number of parts may be expressed, 141, so that a short phrase will describe the leaf completely. 142. All the various terms apply as well to other parts, as to calyx, corolla, petals, &c.

143. Compound Leaves ; Leaflets. 144. The two kinds. 145. Pinnate leaves. 146. Palmate or Digitate. 147. Varieties of pinnate leaves. 148. Number of leaflets. 149. Also of palmate leaves ; why their leaflets are generally fewer than those of pinnate leaves. 150. Twice or thrice compound and decompound leaves.

151. Leaves without distinction of blade and footstalk ; Needle-shaped ; Thread-shaped ; Awl-shaped; Equitant. 152. Stipules; often united with the footstalk, or with each other.

153. The arrangement of leaves on the stem: the three modes, viz. alternate, opposite, whorled.

CHAPTER II.

Section I. — How Propagated from Buds.

154. PLANTS not only grow so as to increase in size or extent, but also multiply, or increase their numbers. This they do at such a rate that almost any species, if favorably situated, and not interfered with by other plants or by animals, would soon cover the whole face of a country adapted to its life.

155. Plants multiply in two distinct ways, namely, by *Buds* and by *Seeds*. All plants propagate by seeds, or by what answer to seeds. Besides this, a great number of plants, mostly perennials, propagate naturally from buds.

156. And almost any kind of plant may be made to propagate from buds, by taking sufficient pains. The gardener multiplies plants artificially in this way,

157. **By Layers and Slips or Cuttings.** In *laying* or *layering*, the gardener bends a branch down to the ground, — sometimes cutting a notch at the bend, or removing a ring of bark, to make it strike root the quicker, — and covers it with earth; then, after it has rooted, he cuts off the connection with the parent stem. Thus he makes artificial stolons (99). Plants which strike root still more readily, such as Willows, he propagates by cuttings or slips, that is, by pieces of stem, containing one or more buds, thrust into the ground or into flower-pots. If kept moist and warm enough, they will generally strike root from the cut end in the ground, and develop a bud above, so forming a new plant out of a piece of an old one. Many woody plants, which will not so readily grow from slips, can often be multiplied

158. **By Grafting or Budding.** In *grafting*, the cutting is inserted into a stem or branch of another plant of the same species, or of some species like it, as of the Pear into the Quince or Apple; where it grows and forms a branch of the *stock* (as the stem used to graft on is called). The piece inserted is called a *scion*. In grafting shrubs and trees it is needful to make the inner bark and the edge of the wood of the scion correspond with these parts in the stock, when they will grow together, and become as completely united as a natural branch is with its parent stem. In *budding* or *inoculating*, a young bud, stripped from one fresh plant, is inserted under the bark of another, usually in summer; there it adheres and gen-

erally remains quiet, as it would have done on the parent bough, until the next spring, when it grows just as if it belonged there.

159. The object of all these ways of artificial propagation from buds is to preserve and to multiply choice varieties of a species which would not be perpetuated from seed. For as the fruit of all the natural branches is alike, so it remains essentially unaltered when borne by branches which are made to grow as artificial branches of another plant, or to take root in the ground as a separate plant. The seeds of an apple or other fruit cannot be depended upon to reproduce the very same sort of apple, — that is, an apple of the very same flavor or goodness. The seeds will always reproduce the same *species*, but not the *individual peculiarities*. These are perpetuated in propagation from buds. This kind of propagation is therefore very important to the cultivator. It takes place naturally in many plants,

160. **By Stolons, Offsets, Runners, or Suckers,** in ways which have already been described (99 to 103, and Fig. 74). These are all forms of natural layering, and they must have taught the gardener his art in this respect. For he merely imitates Nature, or rather sets her at work and hastens her operations. Also,

161. **By Tubers** (74, 75, Fig. 59, 60). These are under-ground branches with lively buds, well charged with prepared nourishment, rendering them more independent and surer to grow. Potatoes and Ground-Artichokes are familiar illustrations of the kind. They are propagated year after year by their buds, or eyes, being very seldom raised from the seed. Each annual crop of tubers is set free at maturity, by the death of all the rest of the plant.

162. **By Corms, Bulbs, and Bulblets**; as explained in paragraphs 77 and 106 to 109. Fig. 76 shows a corm or solid bulb of Crocus, which itself grew by feeding upon its parent, whose exhausted remains are seen underneath : it has already produced a crop of buds, to grow in their turn into another generation of corms, consuming their parent in the process. Bulbs produce a crop of new bulbs from buds in the axils of some of their scales. Tulips, Daffodils, and Garlies propagate very freely in this manner, not only keeping up the succession of generations, but multiplying greatly their numbers.

Analysis of the Section.

154. Plants multiply as well as grow. 155. In two ways; all plants by seeds, many by buds. 156. Most kinds may be propagated by buds artificially. 157. By Layers and Slips or Cuttings. 158. By Grafting or Budding. 159. Object gained by this mode of propagation. 160. It takes place naturally, by Stolons, Offsets, &c. 161. By Tubers. 162. By Corms, Bulbs, and Bulblets.

163. PROPAGATION from buds is really only the division, as it grows, of one plant into two or more, or the separation of shoots from a stock. Propagation from seed is the only true *reproduction*. In the seed an entirely new individual is formed. So the *Seed*, and the *Fruit*, in which the seed is produced, and the *Flower*, which gives rise to the fruit, are the *Organs of Reproduction* (2).

164. Every species at some period or other produces seeds, or something which answers to seeds. Upon this distinction, namely, whether they bear true flowers producing genuine seeds, or produce something merely answering to flowers and seeds, is founded the grand division of all plants into two series or grades, that is, into PHÆNOGAMOUS or FLOWERING PLANTS, and CRYPTOGAMOUS or FLOW-ERLESS PLANTS.

165. **Cryptogamous or Flowerless Plants** do not bear real flowers, having stamens and pistils, nor produce real seeds, or bodies having an embryo ready formed in them. But they produce minute and very simple bodies which answer the purpose of seeds. To distinguish them from true seeds, they are called *Spores*. Ferns, Mosses, Lichens, and Seaweeds, are all flowerless plants, reproduced by spores.

166. **Phænogamous or Flowering Plants** are those which do bear flowers and seeds; the seed essentially consisting of an embryo or germ, ready formed within its coats, which has only to grow and unfold itself to become a plant; as has been fully explained in the first and second sections of Chapter I.

167. Flowerless plants have their organs too minute to be examined without much magnifying, and are too difficult for young beginners. The ordinary or Flowering class of plants will afford them abundant occupation. We are to study first the Flower, then the Fruit and Seed.

SECTION III. — **Flowers.**

§ 1. *Their Arrangement on the Stem.*

168. **Inflorescence** is the term used by botanists for flower-clusters generally, or for the way blossoms are arranged on the stem. Everything about this is governed by a very simple rule, which is this: —

169. Flower-buds appear in the same places that common buds (that is, leaf-buds) do; and they blossom out in the order of their age, the earliest-formed first,

and so on in regular succession.   Now the place for buds is in the axils of the leaves (*axillary buds*, 58), and at the end of the stem (*terminal bud*, 57) : so these are also the places from which flowers spring.   Fig. 138 is a Trillium, with its flower *terminal*, that is, from the summit of the stem. Fig. 139 is a piece of Moneywort, with *axillary* flowers, i. e. from the axils of the leaves.   The Morning-Glory (Fig. 4) also has its flowers axillary.

170. **Solitary Flowers.**   In both these cases the blossoms are solitary, that is, single.   There is only one on the plant in Trillium (Fig. 138).   In Fig. 139, there is only one from the same axil; and although, as the stem grows on, flowers appear in succession, they are so scattered, and so accompanied by leaves, that they cannot be said to form a flower-cluster.

171. **Flower-Clusters** are formed whenever the blossoms are more numerous or closer, and the accompanying leaves are less conspicuous.   Fig. 140 is a cluster (like that of Lily of the Valley, Fig. 3) of the kind called a raceme.   On comparing it with

139
Terminal Flower.

139
Axillary Flowers.

Fig. 139, we may perceive that it differs mainly in having the leaves, one under each blossom-stalk, reduced to little scales, which are inconspicuous.   In both, the flowers really spring from the axils of leaves.   So they do in all the following kinds of flower-clusters, until we reach the Cyme.

172. The leaves of a flower-cluster take the name of *Bracts*.   These are generally very different from the ordinary leaves of the plant, commonly much smaller, and often very small indeed, as in Fig. 140.   In the figures 141 to 144, the bracts are larger, and more leaf-like.   They are the leaves from whose axil the flower arises.   Sometimes there are bracts also on the separate flower-stalks (as on the lower ones in Fig. 140) : to distinguish these we call them *Bractlets*.

173. The flower-stalk or footstalk of a blossom is called a *Peduncle* (96). So the flowers in Fig. 138, 139, &c. are *peduncled* or stalked. But in Fig. 141 they are sitting on the stem, or *sessile*.

174. In clusters we need to distinguish two kinds of flower-stalks; namely, the stalk of the whole cluster, if there be any, and the stalk of each blossom. In such cases we call the stalk of the cluster the *Peduncle*, and the stalk of the individual flowers we name the *Pedicel*. In the Lily of the Valley (Fig. 3, as in Fig. 140), there is the *peduncle* or general flower-stalk (which is here a continuation of the main stem), and then the flowers all have *pedicels* of their own.

175. **Kinds of Flower-Clusters.** Of those which bear their flowers on the sides of a main stalk, in the axils of leaves or bracts, the principal kinds are the *Raceme*, the *Corymb*, the *Umbel*, the *Head*, and the *Spike* with its varieties; also the *Panicle*. In the head and the spike the flowers are sessile. In the others they have pedicels or footstalks of their own.

176. **A Raceme** is a cluster with the blossoms arranged along the sides of a main flower-stalk, or its continuation, and all on pedicels of about the same length. A bunch of Currant-blossoms or berries, or the graceful cluster of the Lily of the Valley (Fig. 3, 140) are good illustrations. Fig. 142 shows the plan of the raceme. Notice that a raceme always blossoms from the bottom to the top, in regular order; because the lower buds are of course the oldest.

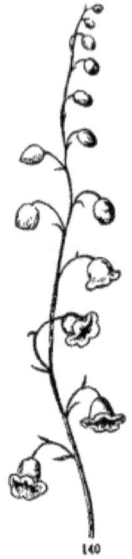

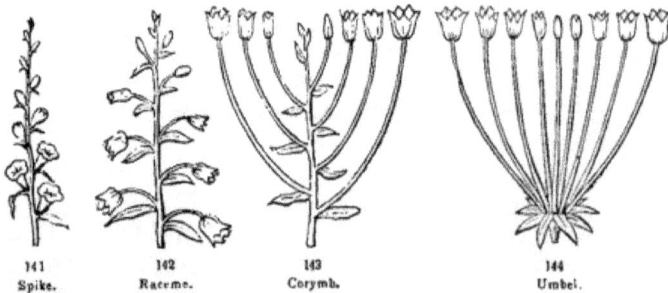

141 142 143   144
Spike. Raceme. Corymb.   Umbel.

177. **A Corymb** is a flat-topped or convex cluster, like that of Hawthorn. Fig.

143 shows the plan of it. It is plainly the same as a raceme with the lower
pedicels much longer than the uppermost.   Shorten the body, or axis, of a corymb
so that it is hardly perceptible, and we
change it into

178. **An Umbel**, as in Fig. 144.   This is a
cluster in which the pedicels all spring from
about the same level, like the *rays* or sticks
of an umbrella, from which it takes its name.
The Milkweed and Primrose bear their
flowers in umbels.

179. The outer blossoms of a corymb or
an umbel plainly answer to the lower blossoms of a raceme.   So the umbel and the
corymb blossom from the circumference
towards the centre, the
outer flower-buds being
the oldest.   By that we
may know such clusters
from cymes.

145
Head.

180. **A Head** is a flower-
cluster with a very short body, or axis, and without any pedicels to the blossoms, or hardly any, so that it has a rounded
form.   The Button-bush (Fig. 145), the Thistle, and the Red
Clover are good examples.

181. It is plain that an umbel would be changed into a head
by shortening its pedicels down to nothing; or, contrarily, that
a head would become an umbel by giving stalks to its flowers.

182. **A Spike** is a lengthened flower-cluster, with no pedicels to
the flowers, or hardly any.   Fig. 141 gives the plan of a spike;
and the common Mullein and the Plantain are good examples.
A head would become a spike by lengthening its axis.   A raceme would become a spike by shortening its pedicels so much
that they could hardly be seen.   The *Catkin* and the *Spadix* are
only sorts of spike.

146.
Catkin.

183. **A Catkin** or **Ament** is a spike with scaly bracts.   The flowers of the Willow, Poplar, Alder, and Birch (Fig. 146) are in catkins.

184. A **Spadix** is a spike with small flowers crowded on a thick and fleshy body or axis. Sweet-Flag and Indian-Turnip are common examples. In Indian-Turnip (Fig. 147) the spadix bears flowers only near the bottom, but is naked and club-shaped above. And it is surrounded by a peculiar leaf or bract in the form of a hood.

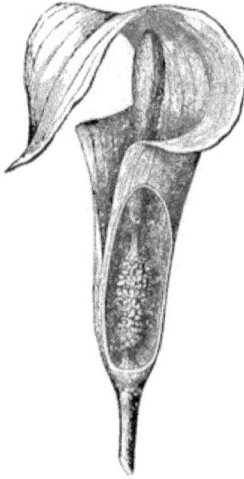

185. Such a bract or leaf enwrapping a spike or cluster of blossoms is named a *Spathe*.

186. A set of bracts around a flower-cluster, such as those around the base of the umbel in Fig. 144, is called an *Involucre*.

187. Any of these clusters may be compound. That is, there may be racemes clustered in racemes, making a compound raceme, or corymbs in corymbs, or umbels in umbels, making a compound umbel, as in Caraway (Fig. 148), Parsnip, Parsley, and all that family. The little umbels of a compound umbel are called *Umbellets;* and their involucre, if they have any, is called an *Involucel*.

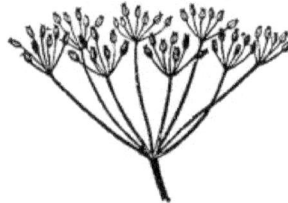

147
Spadix and Spathe.

188. A **Panicle** is an irregularly branching compound flower-cluster, such as would be formed by a raceme with its lower pedicels branched. Fig. 149 shows a simple panicle, the branches, or what would be the pedicels, only once branched. A bunch of Grapes and the flower-cluster of Horsechestnut are more compound panicles. A crowded compound panicle of this sort has been called a *Thyrse*.

148
Compound Umbel.

149
Panicle.

189. A **Cyme** is the general name of flower-clusters of the kind in which a flower always terminates the stem or main peduncle, and each of

its branches. The plan of a cyme is illustrated in the following figures. Fig. 150, to begin with, is a stem terminated by a flower, which plainly comes from a terminal bud or is a *terminal* flower. Fig. 151 is the same, which has started a branch from the axil of each of the uppermost leaves ; each of these ends in a flower-bud. Fig. 152 is the same, with the side branches again branched in the same way, each branch ending in a flower-bud. This makes a cluster looking like a corymb, as

Plan of the Cyme.

shown in Fig. 143 ; but observe that here in the cyme the middle flower, *a*, which ends the main stem, blossoms first ; next, those flowers marked *b* ; then those marked *c*, and so on, the centre one of each set being the earliest ; while in the corymb the blossoming begins with the outermost flowers and proceeds regularly towards the centre. The Elder, the Cornel, and the Hydrangea (Fig. 169) have their blossoms in cymes many times branched in this way ; that is, they have *compound cymes*.

190. A **Fascicle** is only a close or very much crowded cyme, with very short footstalks to the flowers, or none at all, as the flower-cluster of Sweet-William.

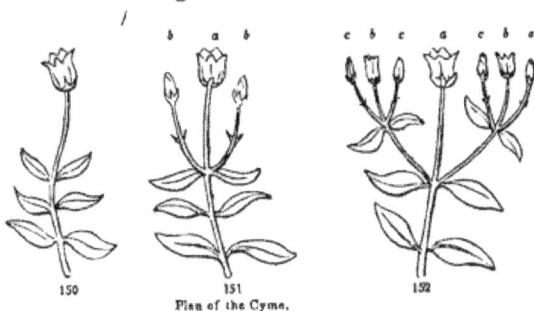

§ 2. *Forms and Kinds of Flowers.*

191. **The Parts of a Flower** were illustrated at the beginning of the book, in Chapter I., Section I. Let us glance at them again, taking a different flower for the example, namely, that of the Three-leaved Stonecrop. Although small, this has all the parts very distinct and regular. Fig. 153 is a moderately enlarged view of one of the middle or earliest flowers of this Stonecrop. (The others are like it, only with their parts in fours instead of fives.) And Fig. 154 shows two parts of each sort, one on each side, more magnified, and separated from the end of the flower-stalk (or *Receptacle*), but standing in their natural position, namely, below or outside a *Sepal*, or leaf of the *Calyx ;* then a *Petal*, or leaf of the *Corolla ;* then a *Stamen ;* then a *Pistil.*

5

192. This is a complete and regular, yet simple flower; and will serve as a pattern, with which a great variety of flowers may be compared.

153

193. When we wish to designate the leaves of the blossom by one word, we call them the *Perianth*. This name is formed of two Greek words meaning "around the flower." It is convenient to use in cases where (as in the Lilies, illustrated on the first page) we are not sure at first view whether the leaves of the flower are calyx or corolla, or both.

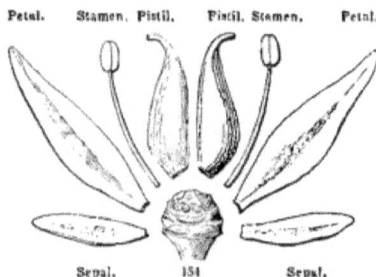

Petal.   Stamen. Pistil.   Pistil. Stamen.   Petal.

Sepal.   154   Sepal.

194. A *Petal* is sometimes to be distinguished into two parts; its *Blade*, like the blade of a leaf, and its *Claw*, which is a kind of tapering base or foot of the blade. More commonly there is only a blade; but the petals of Roses have a very short, narrow base or claw; those of Mustard, a longer one; those of Pinks and the like, a narrow claw, which is generally longer than the blade (Fig. 308).

195. A *Stamen*, as we have already learned (15, 17), generally consists of two parts; its *Filament* and its *Anther*. But the filament is only a kind of footstalk, no more necessary to a stamen than a petiole is to a leaf. It is therefore sometimes very short or wanting; when the anther is *sessile*. The anther is the essential part. Its use, as we know, is to produce pollen.

196. The *Pollen* is the matter, looking like dust, which is shed from the anthers when they open (Fig. 159). Here is a grain of pollen, a single particle of the fine powder shed by the anther of a Mallow, as seen highly magnified. In this plant the grains are beset with bristly points; in many plants they are smooth; and they differ

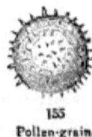

155
Pollen-grain.

greatly in appearance, size, and shape in different species, but are all just alike in the same species; so that the family a plant belongs to can often be told by seeing only a grain of its pollen. The use of the pollen is to lodge on the stigma of the pistil, where it grows in a peculiar way, its inner coat projecting a slender thread which sinks into the pistil, somewhat as a root grows down into the ground, and reaches an ovule in the ovary, causing it in some unknown way to develop an embryo, and thereby become a seed.

197. As to the *Pistil*, we have also learned that it consists of three parts, the *Ovary*, the *Style*, and the *Stigma* (16); that the style is not always present, being only a stalk or support for the stigma. But the two other parts are essential,—the *Stigma* to receive the pollen, and the *Ovary* to contain the ovules, or bodies which are to become seeds. Fig. 156 represents a pistil of Stonecrop, magnified; its stigma (known by the naked roughish surface) at the tip of the style; the style gradually enlarging downwards into the ovary. Here the ovary is cut in two, to show some of the ovules inside. And Fig. 157 shows one of the ovules, or future seeds, still more magnified.

198. **Nature of the Flower.** In the mind of a botanist, who looks at the philosophy of the thing,

A flower answers to a sort of branch. True, a flower does not bear much resemblance to a common branch; but we have seen (90 – 109) what remarkable forms and appearances branches, and the leaves they bear, occasionally take. Flowers come from buds just as branches do, and spring from just the same places that branches do (169). In fact, a flower is a branch intended for a peculiar purpose. While a branch with ordinary leaves is intended for growing, and for collecting from the air and preparing or digesting food,—and while such peculiar branches as tubers, bulbs, &c. are for holding prepared food for future use,— a blossom is a very short and a special sort of branch, intended for the production of seed. If the whole flower answers to a branch, then it follows that (excepting the receptacle, which is a continuation of the flower-stalk) —

The parts of the flower answer to leaves. This is plainly so with the sepals and the petals, which are commonly called the leaves of the blossom. The sepals or calyx-leaves are commonly green and leaf-like, or partly so. And the petals or corolla-leaves are leaves in shape, only more delicate in texture and in color. In many blossoms, and very plainly in a White Water-Lily, the calyx-leaves run into

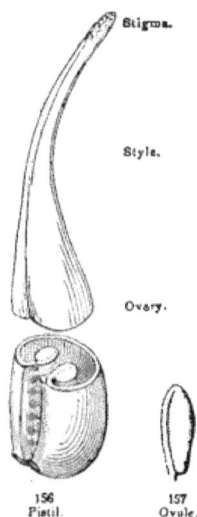

156
Pistil.

157
Ovule.

Stigma.

Style.

Ovary.

corolla-leaves, and the inner corolla-leaves change gradually into stamens, — showing that even stamens answer to leaves.

198ᵃ. How a stamen answers to a leaf, according to the botanist's idea, Fig.
158 is intended to show. The filament or stalk of the stamen answers to the
footstalk of a leaf; and the anther answers to the blade. The lower part of the
figure represents a short filament, bearing an anther which
has its upper half cut away; and the summit of a leaf is
placed above it. Fig. 159 is the whole stamen of a Lily
put beside it for comparison. If the whole anther corre-
sponds with the blade of a leaf, then its two cells, or
halves, answer to the halves of the blade, one on each side
of the midrib; the continuation of the filament, which con-
nects the two cells (called the *connective*), answers to the
midrib; and the anther generally opens along what answer
to the margins of a leaf.

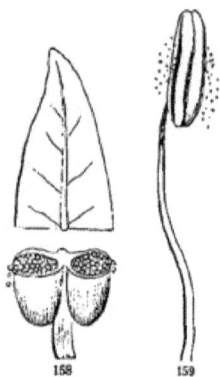

199. It is easy to see how a simple pistil answers to a
leaf. A simple pistil, like one of those of the Stonecrop
(Fig. 154, 156) is regarded by the botanist as if it were
made by the folding up inwards of the blade of a leaf,

158   159
Plan of a Stamen.

(that is, of what would have been a leaf on any branch of the common kind,) so
that the margins come together and join, making a hollow closed bag, which is the
ovary; a tapering summit forms the style, and some part of the
margins of the leaf in this, destitute of skin, becomes the stigma. To understand this better, compare Fig. 160, representing a leaf rolled up in this way, with Fig. 156, and with Fig.
161, which are pistils, cut in two, that the interior of the ovary
may be seen. It is here plain that the ovules or seeds are attached to what answers to the united margins of the leaf. The
particular part or line, or whatever it may be, that the ovules
or seeds are attached to, is called the *Placenta*.

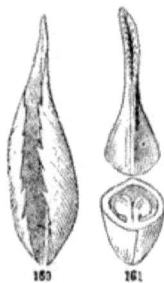

160   161
Plan of Pistil.

200. **Varieties or Sorts of Flowers.** Now that we have learned
how greatly roots, stems, and leaves vary in their forms and
appearances, we should expect flowers to exhibit great variety in different species.
In fact, each class and each family of plants has its flowers upon a plan of its
own. But if students understand the *general plan of flowers*, as seen in the

Morning-Glory, the Lily (Fig. 1 – 12), and the Stonecrop (191), they will soon learn to understand it in any or all of its diverse forms. The principal varieties or special forms that occur among common plants will be described under the families, in the *Flora* which makes the Second Part of this book. There students will learn them in the easiest way, as they happen to meet with them in collecting and analyzing plants. Here we will only notice the leading *Kinds of Variation* in flowers, at the same time explaining some of the terms which are used in describing them.

201. Flowers consist of sepals, petals, stamens, and pistils. There may be few or many of each of these in any particular flower; these parts may be all separate, as they are in the Stonecrop; or they may be grown together, in every degree and in every conceivable way; or any one or more of the parts may be left out, as it were, or wanting altogether in a particular flower. And the parts of the same sort may be all alike, or some may be larger or smaller than the rest, or differently shaped. So that flowers may be classified into several sorts, of which the following are the principal.

202. A **Complete Flower** is one which has all the four parts, namely, calyx, corolla, stamens, and pistils. This is the case in all the flowers we have yet taken for examples; also in Trillium (Fig. 138, reduced in size, and here in Fig. 162, with the blossom of the size of life, and spread open flat).

203. A **Perfect Flower** is one which has both stamens and pistils. A complete flower is of course a perfect one; but many flowers are perfect and not complete; as in Fig. 163, 164.

204. An **Incomplete Flower** is one which wants at least one of the four kinds of organs. This may happen in various ways. It may be

Apetalous ; that is, having no petals. This is the case in Anemony (Fig. 163), and Marsh-Marigold. For these have only one row of flower-leaves, and that is a calyx. The petals which are here wanting appear

162
Complete flower of Trillium.

163
Incomplete flower of Anemony.

in some flowers very much like these, as in Buttercups (Fig. 238) and Goldthread
Or the flower may be still more incomplete, and

Naked, or *Achlamydeous;* that is, without any flower-leaves at all,
neither calyx nor corolla. That is the case in
the Lizard's-Tail (Fig. 164), and in Willows.
Or it may be incomplete by wanting either the
stamens or the pistils; then it is

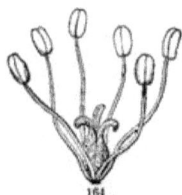

205. **An Imperfect or Separated Flower.** Of course,
if the stamens are wanting in one kind of blossom there must be others that have them. Plants
with imperfect flowers accordingly bear two sorts

164
Flower of Lizard's-Tail.

of blossoms, namely, one sort

Staminate or *Sterile,* those having stamens only, and therefore not
producing seed; and the other

Pistillate or *Fertile,* having a pistil but no good stamens, and ripening seed only when fertilized by pollen
from the sterile flowers. The Oak and Chestnut, Hemp,
Moonseed, and Indian Corn are so. Fig. 165 is one of
the staminate or sterile flowers of Indian Corn; these
form the "tassel" at the top of the stem: their pollen
falls upon the "silk," or styles, of the forming ear below,
consisting of rows of pistillate flowers. Fig. 166 is one
of these, with its very long style. The two kinds of
flowers in this case are

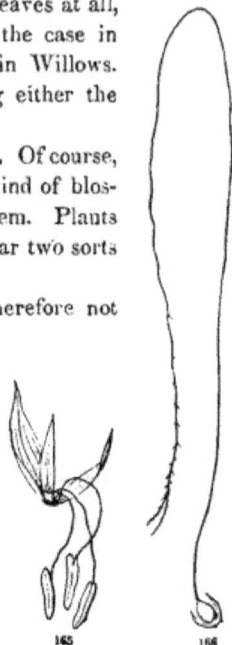

165   166
Indian Corn.

Monœcious; that is, both borne by the same individual plant; as they are also in the Oak,
Chestnut, Birch, &c. In other cases

Diœcious; that is, when one tree or herb
bears flowers with stamens only, and another
flowers with pistils only; as in Willows and
Poplars, Hemp, and Moonseed. Fig. 167 is
a staminate flower from one plant of Moon-
seed, magnified; and Fig. 168, a pistillate flower, borne by a plant from a different
root. There is a third way: some plants produce what are called

167   168
Moonseed Flowers.

Polygamous flowers; that is, having some blossoms with pistils only or with stamens only, and others perfect, having both stamens and pistils, either on the same or on different individuals. The Red Maple is a very good case of this kind; the two or three sorts of flowers looking very differently when they appear in early spring; those of one tree having long red stamens and no good pistil, those of other trees having conspicuous pistils. in some blossoms with no good stamens at all, in others with short ones. There are also what are called *abortive* or

206. **Neutral Flowers**; having neither stamens nor pistils, and so good for nothing except for show. In the Snowball of the gardens and in richly cultivated Hydrangeas all the blossoms are neutral, and no fruit is formed. Even in the wild state of these shrubs, some of the blossoms around the margin of the cluster are neutral (as in the Wild Hydrangea, Fig. 169), consisting only of three or four flower-leaves, very much larger than the small perfect flowers which make up the rest of the cluster. Also what the gardener calls *Double Flowers*, when full, are neutral, as in double Roses and Buttercups. These are blossoms which by cultivation have all their stamens and pistils changed into petals.

207. **A Symmetrical Flower** is one which has an equal number of parts of each kind or in each set or row. This is so in the Stonecrop (Fig. 153), which has five sepals in the calyx, five petals in the corolla, ten stamens (that is, two sets of stamens of five each), and five pistils. Or often it has flowers with four sepals, and then there are only four

petals, eight stamens (twice four), and four pistils. So the flower of Trillium (Fig. 162) is symmetrical; for it consists of three sepals, three petals, six stamens (one

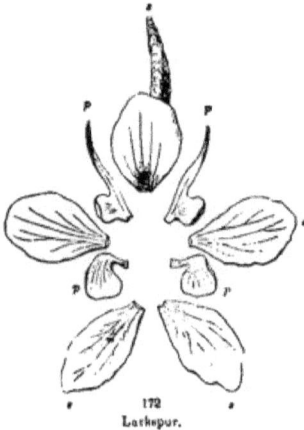

before each sepal and one before each petal), and a pistil plainly composed of three put together, having three styles or stigmas. Flax affords another good illustration of symmetrical flowers (Fig. 170): it has a calyx of five sepals, a corolla of five petals, five stamens, and five styles. In such flowers, and in blossoms generally, the parts *alternate* with each other; that is, the petals stand before the intervals between the sepals, the stamens, when of the same number, before the intervals between the petals, and so on.

208. **An Unsymmetrical Flower** is one in which the different organs or sets do not match in the number of their parts. The flower of Anemony, Fig. 163, is unsymmetrical, having many more stamens and pistils than it has calyx-leaves. And the blossom of Larkspur (Fig. 171) is unsymmetrical, because, while it has five sepals or

172
Larkspur.

173

174
Violet.

leaves in the calyx, there are only four petals or corolla-leaves, but a great many stamens, and only one, two, or three pistils. The sepals and petals are displayed separately in Fig. 172; the five pieces marked *s* are the sepals; the four marked *p* are the petals.

209. **A Regular Flower** is one in which the parts of each sort are all of the same shape and size. The flowers in Flax (Fig. 170) and in all the examples preceding it are regular. While in Larkspur and Monkshood we have not only an *unsymmetrical*, but

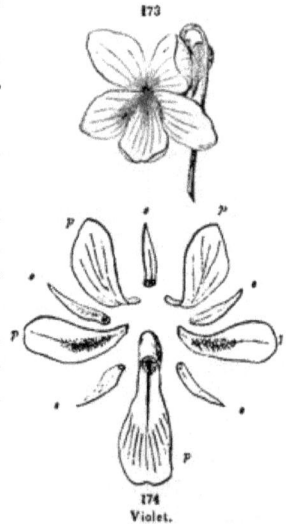

210. **An Irregular Flower**; that is, one in which all the parts of the same sort are not alike. For in the Larkspur-blossom one of the sepals bears a long hollow spur or tail behind, which the four others have not; and the four small petals are of two sorts. The Violet-blossom (Fig. 173) and the Pea-blossom (Fig. 351) are symmetrical (except as to the pistil), but irregular. Fig. 174 shows the calyx and the corolla of the Violet above it displayed; *s*, the five sepals; *p*, the five petals. One of the latter differs from the rest, having a sac or spur at the base, which makes the blossom irregular. So far, most of the examples in this section are from

211. **Flowers with the parts all distinct**, that is, of separate pieces; — the calyx of *distinct sepals*, the corolla of *distinct petals* (i. e. *Polypetalous*), the stamens distinct (separate, &c.), and all the parts growing in regular order out of the receptacle, in other words, *inserted* on the receptacle. These are the simplest or most natural flowers, the parts answering to so many leaves on a short branch. But as in Honeysuckles (Fig. 389) the leaves of the same pair are often found grown together into one, so in blossom-leaves, there are plenty of

212. **Flowers with their parts united or grown together.** The flower of Morning-Glory (Fig. 4) is a good example. Here is the calyx of five separate leaves or sepals (Fig. 176); but in the corolla (Fig. 175) the five petals are completely united into a cup, just as the upper leaves of Honeysuckles are into a round plate. Then, in Stramonium (Fig. 177), the five sepals also are united or grown together almost to their tips into a

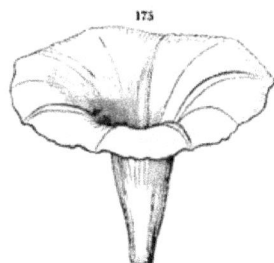

175

176
Morning-Glory.

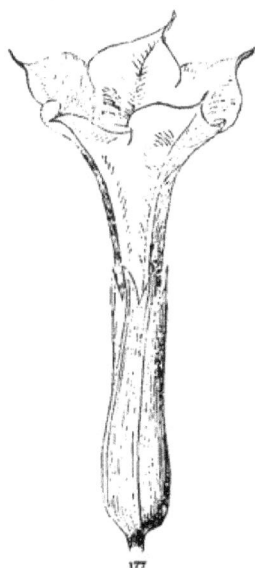

177
Stramonium.

cup or tube; and so are the five petals likewise, but not quite to their tips; and the five teeth or lobes (both of the calyx and of the corolla) plainly show how many leaves there really are in each set. When this is so in the corolla, it forms what is called a

213. **Monopetalous** corolla; i. e. a corolla of one piece. It is so called, whether it makes a cup or tube with the border entire, as in Morning-Glory (Fig. 175), or with the border lobed, that is, the tips of the petals separate, as in Stramonium (Fig. 177), or even if the petals are united only at the bottom, as in the Potato-blossom (Fig. 182). The same may be said of a calyx when the sepals are united into a cup, only this is called *Monosepalous*. A monopetalous corolla (and so of a calyx) is generally distinguishable into two parts, namely, its *Tube* or narrow part below, and its *Border* or *Limb*, the spreading part above. It is *regular* when all sides and lobes of it are alike, as in Fig. 175, 177, &c. It is *irregular* when the sides or parts are different or unequal in size or shape, as in Sage, Dead-Nettle (Fig. 181), the common Honeysuckle, &c. It is

178. Trumpet-Honeysuckle.

Tubular, when long and narrow without a conspicuous border, as in Fig. 178, or

Trumpet-shaped ; tubular, gradually enlarging upwards, as in Trumpet-Creeper and Trumpet-Honeysuckle (Fig. 178) ;

Funnel-shaped or *Funnel-form* (like a funnel or tunnel) ; when the tube opens gradually into a spreading border, as in Morning-Glory (Fig. 175) and Stramonium (Fig. 177) ;

Bell-shaped or *Campanulate ;* when the tube is wide for its length and the border a little spreading, like a bell, as in Harebell (Fig. 179).

179   180   181

Salver-shaped ; when a slender tube spreads suddenly into a flat border, as in Phlox (Fig. 180).

Wheel-shaped or *Rotate ;* same as salver-shaped, with the tube very short or none, as in the corolla of the Potato (Fig. 182) and the Nightshade (Fig. 183).

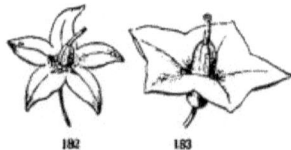

Labiate or *Two-lipped ;* when the border divides into two parts, or *lips*, an upper and a lower (sometimes likened to those of an animal with gaping mouth), as in Sage, Dead-Nettle (Fig. 181, and the like. This is one of the irregular forms of monopetalous corolla, and the commonest.

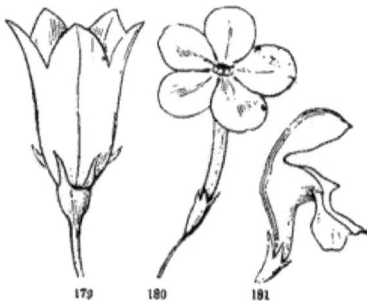

182   183

214. Stamens united are also common. They may be united by their filaments or by their anthers. In the Cardinal-flower (Fig. 184), and other Lobelias, both the anthers (*a*) and the filaments (*f*) are united into a tube. So also in the Pumpkin and Squash. Botanists use the following terms to express the different ways in which stamens may be connected. They are

Syngenesious, when the anthers are united into a ring or tube, as in Lobelia (Fig. 184 *a*), and in the Sunflower, and all that family.

Monadelphous (i. e. in one brotherhood), when the filaments are united all into one set or tube, as in Lobelia (Fig. 184 *f*), and the Mallow Family (Fig. 185); also in Passion-flowers and Lupines (Fig. 187).

184. Lobelia.

Diadelphous (in two brotherhoods), when the filaments are united in two sets. Fig. 186 shows this in the Pea, and the like, where nine stamens are combined in one set and one stamen is left for the other.

185. Mallow.

Triadelphous (in three brotherhoods), when the filaments are united or collected in three sets, as in the Common St. John's-wort or Hypericum (Fig. 297); and

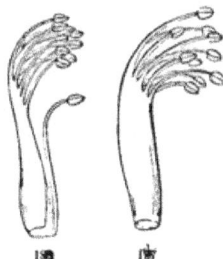

Polyadelphous (in many brotherhoods), when combined in more than three sets, as in some St. John's-worts.

215. Pistils united are very common. Two, three, four, or more grow together at the time of their formation, and make a *Compound Pistil.* Indeed, wherever there is a single pistil to a flower, it is much oftener a compound pistil than a simple one. But, of course, when the pistils of a flower are more than one, they are all simple. Pistils may be united in every degree, and by their ovaries only, by their styles only (as they are slightly in Prickly-Ash), or even by their stigmas only (as in Milkweeds), or by all three. But more commonly the ovaries are united into one *Compound Ovary,* while the styles or stigmas are partly separate or *distinct.* Three degrees of union are shown in these figures. Fig. 188, two pistils of a Saxi-frage, their ovaries united only part way up (cut across both above and below).

Fig. 189, pistil of Common St. John's-wort, plainly composed of three simple ones, with their ovaries completely united, while their slender styles are separate. Fig. 190, same of Shrubby St. John's-wort, like the last, but with the three styles also grown together into one, the little stigmas only separate; but as it gets older this style generally splits down into three, and when the pod is ripe it also splits into three, plainly showing that this compound pistil consists of three united into one. On turning now to Fig. 8 and Fig. 10 to 12 on the same page, it will be seen that the pistil in Morning-Glory and in Lily is a compound one, made of three

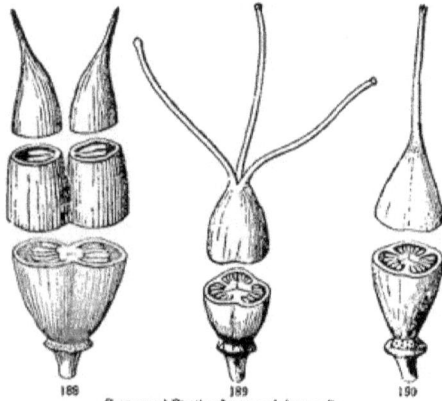

Compound Pistils of two and three cells.

united even to their stigmas. This is shown externally, by the stigma being somewhat three-lobed in both. And it becomes perfectly evident on cutting the ovary in two, bringing to view the three *cells* (Fig. 12, as in Fig. 189, 190), each answering to one simple ovary.

216. So compound ovaries generally have as many cells as there are simple pistils or pistil-leaves in their composition; and have the *placentas* (199) bearing the seeds all joined in the centre: that is, the *placentas* or compound *placenta in the axis*. But sometimes the *partitions* or divisions between the cells vanish, as in Pinks: then the compound pistil is only one-celled. And sometimes there never were any partitions; but the pistil was formed of two, three, or more open pistil-leaves grown together from the first by their edges, just as petals join to make a monopetalous corolla. Then the ovules or seeds, or the *placentas* that bear them, are *parietal*, that is, are borne on the *parietes* or wall of the ovary. Fig. 191 is the lower part of a compound ovary, with three *parietal placentas* or seed-bearing lines; and Fig. 192 is

One-celled compound ovary, with placentas parietal.

a diagram, to explain how such a pistil is supposed to be made of three leaves united by their edges, and these edges bearing the ovules or seeds.

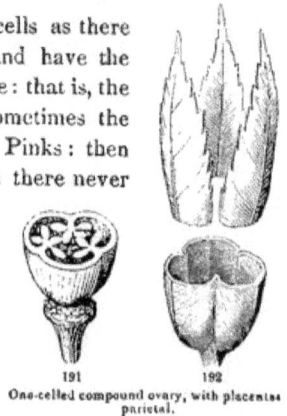

217. Flowers with one set of Organs united with another. The natural way is, for all the parts to stand on the receptacle or end of the flower-stalk, — the stem-part of the blossom (191). Then the parts are said to be *free*, or to be *inserted on the receptacle*. So it is in the Buttercup, Lily, Trillium (Fig. 162), Flax, &c. But in many flowers one set of organs grows fast to another set, or, as we say, is inserted on it. For instance, we may have the *Petals and Stamens inserted on the Calyx*, as in the Cherry and all the Rose family. Fig. 193 is a flower of a Cherry, cut through the middle lengthwise, to show the petals and stamens growing on the tube or cup

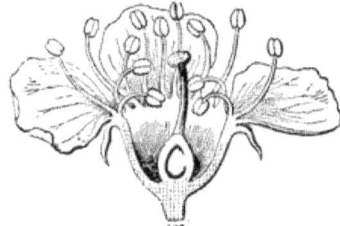
193
Half of a Cherry-blossom.

of the calyx. The meaning of it is that all these parts have grown together from their earliest formation. Next we may have the

Calyx cohering or grown fast to the Ovary, or at least its cup or lower part grown to the ovary, and forming a part of the thickness of its walls, as in the Currant and Gooseberry, the Apple and Hawthorn. Fig. 194 is a flower of Hawthorn cut through lengthwise to show this. In such cases

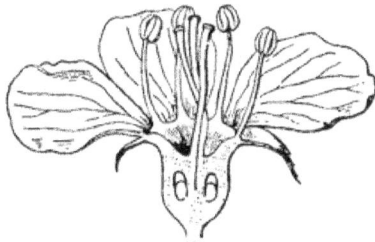
191
Half of a Hawthorn-blossom.

all other parts of the blossom appear to grow on the ovary. So the ovary is said to be *inferior*, or, which is the same thing, the calyx (i. e. its lobes or border) and the rest of the blossom, *superior*. Or else we say "*calyx coherent with the ovary*," which is best, because it explains the thing.

Stamens inserted on the Corolla. The stamens and the corolla generally go together. And when the corolla is of one piece (i. e. *monopetalous*, 213), the stamens almost always adhere to it within, more or less; that is, are borne or "inserted on the

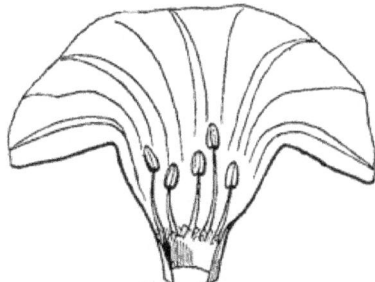
195. Morning-Glory.

corolla." Fig. 195 is the corolla of Morning-Glory laid open, to show the stamens inserted on it, i. e. grown fast to it, towards the bottom. We may even have the

Stamens inserted on the Style, that is, united with it even up to the stigma. It is so in the Orchis family.

218. **Gymnospermous or Open and Naked-seeded Pistils.** This is the very peculiar pistil which belongs to Pines, Spruces, Cedars, and all that family of plants; and it is the simplest of all. For here the pistil is an open leaf or scale, bearing two or three ovules on its upper or inner surface. Each scale of a Pine-cone is an open pistil, and the ovules, instead of being enclosed in an ovary which forms a pod, are

naked, and exposed to the pollen shed by the stamen-bearing flowers, which falls directly upon them. Fig. 196 is a view of the upper side of an open pistil or scale from a forming Larch-cone, at flowering-time, showing the two ovules borne on the face of it, one on each side near the bottom. Fig. 197 is the same grown larger, the ovules becoming seeds. When ripe and dry, the scales turn back, and the naked seeds peel off and fall away.

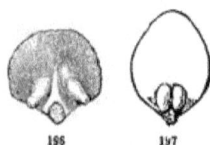

219. Plants which have such open scales for pistils accordingly take the name of GYMNOSPERMOUS or *Naked-seeded.* The Pine family is the principal example of the kind (see p. 201). All other Flowering plants are

ANGIOSPERMOUS, that is, have their ovules and seeds produced in a seed-vessel of some sort.

Analysis of the Section.

168. Arrangement of Flowers, or Inflorescence. 169. Situation of Flower-buds : terminal and axillary. 170. Solitary flowers. 171. Flower-clusters. 172. Bracts and Bractlets. 173, 174. Flower-stalks: Peduncle and Pedicels. 175. Kinds of flower-clusters. 176. Raceme; order of opening of the blossoms. 177. Corymb. 178. Umbel. 179. Comparison with Raceme, &c. 180. Head. 181. Comparison with the Umbel, and, 182. the Spike. 183. Catkin or Ament. 184. Spadix. 185. Its Spathe. 186. Involucre. 187. Compound Clusters: Umbellets; Involucel. 188. Panicle; Thyrse. 189. Cyme. 190. Fascicle.

191. Flowers: their parts illustrated by the Stonecrop: 192. A pattern flower. 193. Leaves of flower or Perianth. 194. Petal; its Blade and Claw. 195. Stamen; its parts. 196. Pollen ; its structure and use. 197. Pistil ; its parts. 198. Nature of the flower; its parts answer to leaves. 198ª. How a stamen answers to a leaf. 199. How a pistil answers to a leaf : Placenta.

200. Sorts of Flowers : one general plan : 201. Varied in several ways. 202. Complete flower. 203. Perfect flower. 204. Incomplete flower: apetalous; naked. 205. Imperfect or separated flowers: staminate or sterile ; pistillate or fertile ; monœcious, diœcious, or polygamous. 206. Neutral flowers.

207. Symmetrical flowers. 208. Unsymmetrical flowers. 209. Regular flowers. 210. Irregular flowers.

211. Flowers with the parts distinct. 212. With their parts grown together. 213. Monopetalous corolla, &c.: its varieties in form. 214. Stamens united; syngenesious, monadelphous, diadelphous, triadelphous, and polyadelphous. 215. Pistils united into a Compound Pistil: illustrations. 216. Those with two or more cells and placentas in the centre; of one cell with placentas parietal or on the walls.

217. Flowers with one set of organs united with another; as petals and stamens with the calyx; the tube or cup of the calyx with the ovary; stamens with the corolla; or with the style.

218. Gymnospermous or Naked-seeded Pistil of Pines, &c. 219. Division of plants on this account.

SECTION IV. — **Fruit and Seed.**

§ 1. *Seed-Vessels.*

220. AFTER the flower comes the Fruit. The ovary of the flower becomes the *Seed-vessel* (or *Pericarp*) in the fruit. The ovules are now seeds.

221. A **Simple Fruit** is a seed-vessel formed by the ripening of one pistil (with whatever may have grown fast to it in the flower, such as the tube of the calyx in many cases, 217). Simple fruits may be most conveniently classified into *Fleshy Fruits, Stone Fruits,* and *Dry Fruits.*

222. The principal sorts of fleshy fruits are the *Berry,* the *Pepo,* and the *Pome.*

223. A **Berry** is fleshy or pulpy throughout. Grapes, tomatoes, gooseberries, currants, and cranberries are good examples. (Fig. 198 shows a cranberry cut in two.) Oranges and lemons are only a kind of berry with a thicker and leathery rind.

224. The **Pepo** or **Gourd Fruit** (such as a squash, melon, cucumber, and bottle-gourd, Fig. 199) is only a sort of berry with a harder rind.

198. Berry.

199. Pepo.

225. A **Pome** or **Apple-Fruit** is the well-known fruit of the Apple, Pear, Quince, and Hawthorn. It comes from a compound pistil with a coherent calyx-tube (that is, from such a flower as Fig. 194), and this calyx,

growing very thick and fleshy, makes the whole eatable part or flesh of the fruit in the haw and the quince.   The real seed-vessels in the quince (Fig. 201), apple (Fig. 200), and the like, consist of the five thin, parchment-like cells of the core, containing the seeds.   In the quince, all the flesh is calyx.   But in the pear and apple the flesh of the core, viz. all inside of the circle of greenish dots which are seen on cutting the apple across (Fig. 200), belongs to the receptacle of the flower, which here rises so as to surround the real seed-vessels. Cutting the apple lengthwise, these dots come to view as slender greenish lines, separating what belongs to the core from what belongs to the calyx : they are the vessels which in the blossom belong to the petals and the stamens above.   In the haw, the cells become thick and stony, and so form a kind of

226. **Stone-Fruit** or **Drupe.**   Plums, cherries, and peaches (Fig. 202) are the commonest and best examples of the stone-fruit.   It is a fruit in which the outer part becomes fleshy or pulpy, like a berry, while the inner part becomes hard or stony, like a nut.   So the *Stone* (or *Putamen*, as the botanist terms it) does not belong to the seed, but to the fruit.   It has the seed in it, with coats of its own.

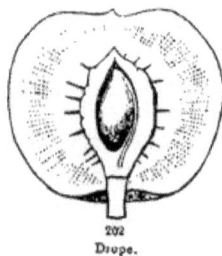

201
Pome.

227. **Dry Fruits** are those that ripen without flesh or pulp.   They are either dehiscent or indehiscent.   Dehiscent seed-vessels are those which split or burst open, in some regular way, to discharge the seeds.   *Indehiscent* seed-vessels are those that remain closed, retaining the seed until they grow, or until the seed-vessel decays.   All stone fruits and fleshy fruits are of course indehiscent.

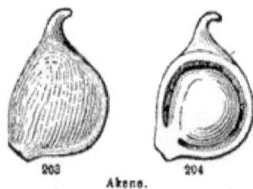

202
Drupe.

228.   The sorts of *indehiscent* dry fruits that we need to distinguish are the *Akene*, the *Grain*, the *Nut*, and the *Key.*

229. **The Akene** includes all dry, one-seeded, closed, small fruits, such as are generally mistaken for naked seeds; such, for instance, as the little seed-like fruits of Buttercups. (Fig. 203 is one of these, whole, a good deal enlarged; Fig. 204, one with part of the wall cut away.)

203   204
Akene.

That they are not seeds is plain from the way they are produced, and from their bearing a style or stigma, at least when young. They are evidently pistils ripened; and on cutting them open, the seed is found whole within (Fig. 204).

230. A **Grain** (or *Caryopsis*) is the same as an akene, except that the thin seed-vessel adheres firmly to the whole surface of the seed. Indian corn, wheat, rye, and all such kinds of grain are examples.

231. A **Nut** is a hard-shelled, one-seeded, indehiscent fruit, like an akene, but on a larger scale. Beechnuts, chestnuts, and acorns (Fig. 205) are familiar examples. In all these the nut is surrounded by a kind of involucre, called a *Cupule* or *Cup*, which, however, is no part of the fruit. In the Oak, the cupule is a scaly cup; in the Beech and Chestnut, it is a kind of bur; in the Hazel, it is a leaf-like cup or covering; in Hop-Hornbeam, it is a thin and closed bag. The fruit of the Walnut, Butternut, and the like, is between a drupe and a nut, having a fleshy outer layer.

205
Nut and Cupule.

232. A **Key** or **Key-Fruit** (called by botanists a *Samara*) is like an akene or nut, or any other indehiscent, one-seeded fruit, only it is winged. The fruits of the Ash (Fig. 206) and of the Elm (Fig. 207) are of this kind. That of the Maple consists of two keys partly joined at the base, both from one flower (Fig. 208).

233. *Dehiscent* Fruits, or dry fruits which split or burst open in some regular way, take the general name of

234. **Pods.** These generally split lengthwise when ripe and dry. Pods formed of a simple pistil mostly open down their inner edge, namely, that which answers to the united margins of the pistil-leaf. Compare Fig. 160 with Fig. 209: the latter is the simple pod of a Marsh-Marigold open after ripening, and the seeds fallen, so becoming a leaf again, as it were. Some such pods also split down the back as well as down the inner side; that

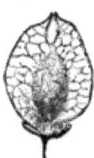
207
Key.

208
Pair of Keys.

206

209
Opened Follicle.

is, along what answers to the midrib of the leaf; as do pea-pods (Fig. 211).

6

235. A **Follicle** is such a simple pod which opens down one side only. The pods of Peony, Columbine, and Marsh-Marigold (Fig. 210) are follicles.

236. A **Legume** is a pod of a simple pistil, which splits into two

210
Follicle.

pieces. It is the fruit of the Pea or Pulse family. Fig. 211 is a legume of the Pea, open, separated into its two valves.

237. A **Capsule** is the pod of any compound pistil. When capsules open regularly, they either split *through the partitions*, or where these would be, as in the pod of St. John's-wort (Fig. 212); this divides them into so many follicles, as it were, which open down the inner edge: or else they split open *into the back of the cells*, as in the pods of the Lily, the Iris (Fig. 213), &c.

211
Legume.

238. The pieces into which a pod splits are called its *Valves*. So a follicle (Fig. 210) is *one-valved*; a legume (Fig. 211), *two-valved*; the capsules in Fig. 212 and 213, both *three-valved*, &c.

239. Two or three forms of capsule have peculiar names. The principal sorts are the *Silique*, the *Silicle*, and the *Pyxis*.

240. A **Silique** (Fig. 214) is the pod of the Cress family. It is slender, and splits into two valves or pieces, leaving behind a partition in a frame which bears the seeds.

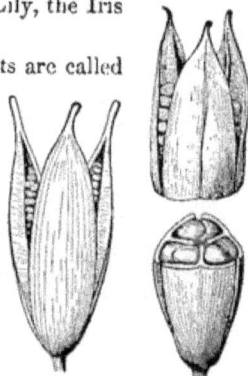

212   213
Capsules, opening.

241. A **Silicle** or **Pouch** is only a silique not much longer than broad. Fig. 215 is the silicle of Shepherd's Purse; Fig. 216, the same with one valve fallen.

242. A **Pyxis** is a pod which opens crosswise, the top separating as a lid. Fig. 217 shows it in the Common Purslane; the lid falling off.

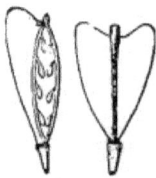

214
Silique.

216   215
Silicle.

217. Pyxis.

243. There remain a few sorts of fruits which are more or less compound or complex. They may be classed under the heads of *Aggregated*, *Accessory*, and *Multiple* fruits.

244. **Aggregated Fruits** are close clusters of simple fruits all of the same flower. The raspberry and the blackberry are good examples. In these, each grain is a *drupelet* or stone-fruit, like a cherry or peach on a very small scale.

245. **Accessory Fruits** are those in which the flesh or conspicuous part belongs to some accessory (i. e. added or altered) part, separate from the seed-vessel. So that what we eat as the fruit is not the fruit at all in a strict botanical sense, but a calyx, receptacle, or something else which surrounds or accompanies it. Our common checkerberry is a simple illustration. Here the so-called berry is a free or separate calyx, which after flowering becomes thick and fleshy, and encloses the true seed-vessel, as a small pod within. Fig. 218 shows the young pod, partly covered by the loose calyx.

 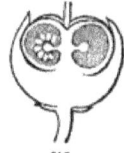

Fig. 219 is the ripe checkerberry, cut through the middle lengthwise, the calyx now thick, juicy, and eatable, and enlarged so as to enclose the small, dry pod.

246. A *Rose-hip* (Fig. 220) is a kind of accessory fruit, looking like a pear or a haw. But it consists of the tube of the calyx, lined by a hollow receptacle, which bears the real fruits, or seed-vessels, in the form of bony akenes. Fig. 221, a rose-hip when in flower, cut through lengthwise, shows the whole plan of it: the pistils are seen attached to the sides of the urn-shaped receptacle, and their styles, tipped with the stigmas, project a little from the cavity, near the stamens, which are borne on the rim of the deep cup.

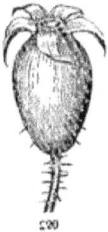 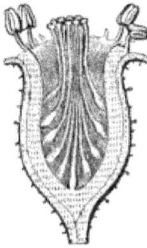

Rose-hip.

247. A *Strawberry* is an accessory fruit of a different shape. Fig. 222 is a forming one, at flowering time, divided lengthwise: below is a part of the calyx; above this, a large oval or conical receptacle, its whole surface covered with little pistils. In ripening this grows

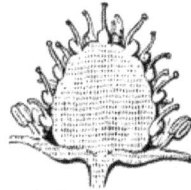

222. Young Strawberry.

vastly larger, and becomes juicy and delicious. So that, in fact, what is called a berry is only the receptacle of the flower, or the end of the flower-stalk, grown very large and juicy, and not a seed-vessel at all, but bearing plenty of one-seeded dry seed-vessels (akenes, 229), so small that they are mistaken for seeds.

248. **Multiple Fruits** are masses of simple or accessory fruits belonging to different flowers, all compacted together. *Mulberries* (Fig. 223) are of this sort. They

223
Mulberry.

look like blackberries, but each grain belongs to a separate flower; and the eatable pulp is not even the seed-vessel of that, but is a loose calyx grown pulpy, just like that of Checkerberry, and surrounding an akene, which is generally taken for a seed. The pine-apple is much like a mulberry on a large scale. A fig is a multiple fruit, being a hollow flower-stalk grown pulpy, the inside lined by a great number of minute flowers.

249. So, under the name of fruit very different things are eaten. In figs it is a hollow flower-stalk; in pine-apples and mulberries, clusters of flower-leaves, as well as the stalk they cover; in strawberries, the receptacle of a flower; in blackberries, the same, though smaller, and a cluster of little stone-fruits that cover it; in raspberries, the little stone-fruits in a cluster, without the receptacle. In checkerberries, quinces, and (as to all but the core) apples and pears, we eat a fleshy enlarged calyx; in peaches and other stone-fruits, the outer part of a seed-vessel; in grapes, gooseberries, blueberries, and cranberries, the whole seed-vessel, grown rich and pulpy.

225　　226

250. The Cone of Pine (Fig. 224) and the like is a sort of multiple fruit. Each scale is a whole pistillate flower, consisting of an open pistil leaf, ripened, and

224
Pitch-pine Cone.

bearing on its upper face one or two naked seeds,—as explained at the end of the last section (218, 219). Fig. 225 shows the upper side of one of the thick scales taken off, bearing one seed; the other, removed, is shown, with its wing, in Fig. 226.

§ 2. *Seeds.*

252. A **Seed** is an ovule fertilized and matured, and with a germ or embryo formed in it.

253. In the account of the growth of plants from the seed, at the beginning of the book (Chapter I. Section I.), seeds have already been considered sufficiently

for our purpose. As the pupil advances farther in his botanical studies, he will learn much more about them, as well as about fruits and flowers, in the Lessons in Botany, and other works.

254. A seed consists of its *Coats* and its *Kernel.* Besides the true seed-coats, which are those of the ovule, an outer loose covering, generally an imperfect one, is occasionally superadded while the seed is growing. This is called an *Aril.* Mace is the aril of the nutmeg. The scarlet pulpy covering of the seeds of the Strawberry-tree and the Staff-tree or Waxwork is also an aril.

255. **The Seed-Coats** are commonly two, an outer and an inner; the latter generally thin and delicate. The outer coat is sometimes close and even, and fitted to the kernel, as in Morning-Glory (Fig. 227); sometimes it is furnished with a tuft of long hairs, as in Milkweed (Fig. 229), or else is covered with long woolly hairs, as in the Cotton-plant, where they form that most useful material, Cotton-wool. In some cases the outer coat is extended into a thin border or wing, as in the Trumpet-Creeper (Fig. 228). Catalpa-seeds have a fringe-like wing or tuft at each end. The seeds of Pines are winged at one end (Fig. 226). All these tufts and wings are contrivances for rendering such seeds buoyant, so that, when shed, they may be dispersed by the wind. Thistle-down, and the like, is a similar contrivance on the fruit or akene.

256. The seed is often supported by a stalk of its own, the *Seed-stalk.* Where the seed separates, it leaves a mark, called the *Scar* or *Hilum.* This is conspicuous in a bean and a pea, and is remarkably large in a horsechestnut.

257. **The Kernel** is the whole body of the seed within the coats. It consists of the *Embryo,* and of the *Albumen,* when there is any.

258. **The Albumen** is a stock of prepared food, for the embryo to live upon at the outset, in those cases where it has not a similar supply laid up in its cotyledons (32–35, 45). In Fig. 17, 44, and 49, the seeds have albumen. In Fig. 25, 32, 40, and 42, they have none, but the whole kernel consists of

259. **The Embryo,** or rudimentary plantlet in miniature, the body in the seed which grows. To this the seed, the fruit, and the blossom are all subservient. The albumen of the seed, when there is any, is intended to nourish the embryo when it

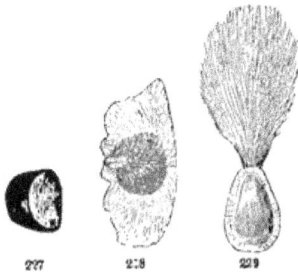

grows, until it can provide for itself; the seed-coats to protect it, especially after it is shed; the seed-vessel, to protect it and to nourish it while forming; the stamens and pistil, to originate it.

260. The embryo consists of its *Radicle* or original stemlet, from one end of which the root starts, from the other the stem is continued; also of one or more *Cotyledons* or *Seed-leaves*, and often of a *Plumule* or bud for continuing the stem upwards. How the embryo grows into a plant, was fully explained at the commencement of this book.

Analysis of the Section.

§ 1. TABLE OF KINDS OF FRUIT.

		Parag.
Fruits are	1. SIMPLE: seed-vessels of one pistil, ...	221
	2. AGGREGATED: clusters of seed-vessels all of the same flower, .	244
	3. ACCESSORY: the flesh, &c. external to and separate from the pistil,	245-247
	4. MULTIPLE: composed of the simple or accessory fruits of more than one flower, ...	248

Simple Fruits are	Flesby Fruits (222), such as the		BERRY,	223
			PEPO,	224
			POME,	225
	Stone-Fruits, or the .		DRUPE,	226
	Dry Fruits, 227,	Indehiscent, 228,	AKENE,	229
			GRAIN,	230
			NUT,	231
			KEY,	232
		Dehiscent, or Pods, 233,	Of a Simple Pistil,	FOLLICLE,235
				LEGUME, 236
			of a Compound Pistil,	CAPSULE, 237
				SILIQUE, 240
				SILICLE, 241
				PYXIS, 242

Multiple Fruits are	Angiospermous, or closed, 219, 248.		
	Gymnospermous, or naked-seeded. 218, 219,	. . .	CONE, 250

§ 2. SEEDS.—252. What a seed is. 253. Its nature already considered. 254. Its parts; Aril occasionally met with. 255. Its coats, and the appendages, wings, &c. 256. Seed-stalk, and scar. 257. Kernel. 258. Albumen, sometimes present; its office. 259. Embryo, to which all the other parts of the seed, the fruit, and the flower are subservient. 260. Parts of the Embryo: Radicle, or Stemlet; Cotyledons, or Seed-leaves; Plumule, or Bud.

CHAPTER III.

WHY PLANTS GROW, WHAT THEY ARE MADE FOR, AND WHAT THEY DO

261. We have now become acquainted with all the organs of plants, both those concerned in their life and growth, or *vegetation*, and those concerned in multiplying their numbers, that is, in *reproduction*. The first being the root, stem, and leaves; the second, the flowers (essentially the stamens and pistils), with their result, fruit and seed. We have learned, also, how plants grow from the seed, produce part after part, branch after branch, and leaf after leaf, and at length blossom and go to seed. We see that plants, with their organs, that is, *instruments*, are a kind of living machines at work; and it is now time to ask, *How they operate, What they bring to pass*, and *What is the object or the result* of their doings. Such questions as these, young people, with their curiosity awakened, would be likely to ask, and they ought to be answered. To understand these things completely, one must know something of chemistry and vegetable anatomy,* — which we do not propose here to teach. But a general account of the matter may be given in a simple way, which shall be perfectly intelligible, and may give a clear idea of the purpose which plants were created to fulfil in the world, and how they do it. Let us begin by considering

262. **The Plant in Action.** Take any living plant, — it matters not what one, — and consider what it is doing. For greater simplicity, take some young plant or seedling, where vegetation goes on just as in a full-grown herb or tree, only on a smaller scale. The plant is

263. **Absorbing,** or drawing in what it lives upon, from the soil and the air. This is moisture, air, and other matters which the rain, as it soaks into the ground, may have dissolved on its way to the roots. It is by the roots, lodged in the damp soil, that most of the moisture which plants feed upon is taken in, and with this they always get some earthy matter. This earthy matter makes the ashes which are left after burning a piece of wood, a leaf, or any part of a plant. Moisture is

---

* After studying this chapter, the pupil will be ready to learn more of the subject in the *Lessons in Botany and Vegetable Physiology*. Lessons 22, 23, 24, and 25 treat of Vegetable Anatomy; and Lesson 26, of the Plant doing its work.

also absorbed by the leaves, either from drops of rain or dew, or from the vapor of water in the air. Air is largely absorbed by the leaves, and some also by the roots, either as dissolved in water, or else directly from the crevices and pores of the soil, which are filled with air.

264. *Plants absorb their food by their surface.* Animals have an internal cavity, — a *stomach*, — to hold their food; and from the stomach it is taken into the system. Plants have nothing of this kind. They absorb their food by their surface, — by the skin, as it were; and when very young and with the whole surface fresh and thin, by one part almost as much as another. But as they grow older and the skin hardens, they absorb mostly by their fresh rootlets and the tips of the roots, and by the leaves, — the former spread out in the soil, the latter spread out in the air. For while the skin or bark of the older parts of the roots is hardening, new tips and rootlets are always forming in growing plants, with a fresh surface, which absorbs freely. And as to the leaves, they are renewed every year (even evergreens produce a new crop annually, and the old ones fall after a year or two); and the skin of every leaf, especially that of the under side, is riddled with thousands of holes or little mouths (called *Breathing-pores*), which open into the chambers or winding passages of the pulp of the leaf, so that the air may circulate freely throughout the whole.

265. *Plants absorb their food all in the fluid form.* They are unable to take in anything in a solid state. They *imbibe* or drink in all their food, in the form of water, with whatever the water has dissolved, and of air or vapor, by one or both of which their leaves and roots are surrounded. The reason they imbibe only fluid is this. The roots, leaves, and all the rest of the plant, under the microscope, are seen to be made up of millions of separate little cavities, each cut off from the surrounding ones by closed partitions of membrane. All that the plants take into their system has to pass through these partitions of membrane, — which fluid (air or moisture) alone can do.

266. *The common juices of plants are called Sap.* What they take in from the soil and the air, not being digested or made into vegetable matter, is called *Crude Sap*. All that the roots imbibe has to be carried up to the leaves to be digested there. So while the roots are absorbing, the stem is

267. **Conveying the Crude Sap to the Leaves.** There is no separate set of vessels, and no open tubes or pipes for the sap to rise through in an unbroken stream, in the way people generally suppose. The stem is made up, like the root, of cavities,

or cells divided off by whole partitions; and to rise an inch the sap generally has to pass through several hundred such partitions. When there is much wood, the sap rises mostly through that. Now the fibres and the vessels of the wood are tubes, most of them several times longer than wide; but their ends do not open into each other; a closed partition divides each cavity from the next, which the sap has to get through some way or other. How it gets through so readily, we do not altogether know; but there is no doubt about the fact.

268. Carried into the leaves, and distributed through their broad surface, the crude sap is exposed to the light and air. A large part of it is water; and each drop of this serves to bring up a minute portion of earthy matter, which it dissolved out of the soil. Most of the water, no longer wanted, is evaporated from the leaves by the warmth of the sun, and *exhaled;* that is, it escapes in vapor into the air, mostly through the breathing-pores (264). What remains, the plant is at the same time

269. **Digesting or Assimilating.** Assimilating is the proper word. To assimilate is to make similar, or to turn into its own substance. This is just what plants do in their leaves. They change into vegetable matter that which was mineral matter (air, earth, or water) before. This they do only in the leaves, or other green parts, and in these only when they are exposed to the light of day, that is, to the influence of the sun. We see, therefore, why plants are so dependent on the light. They cannot grow without it, except so far as they are fed by vegetable matter prepared beforehand;— as the seedling is fed at the beginning, by vegetable matter of the parent plant stored up in the seed (Chap. II. Sect. II.); and potato-shoots, by that provided in the tuber or potato (74, 75), &c. This enables them to begin their growth in the dark. But the inheritance only serves to set up the young plants; when they have exhausted it, they have to work for themselves, to take in air and water, and a little earth, and *assimilate* it,— i. e. make vegetable matter of it,— in their leaves or other green parts, with the help of sunshine. This they do throughout the whole growing season.

270. The new-made vegetable matter is dissolved in the water or the sap in the leaf, and forms a thin mucilage. This is prepared or *Elaborated Sap,* fit to be used in growth; for it contains the same material as that which the plant itself is built of. It is to the plant just what the prepared clay is to the earthen vessel, or to the bricks of which the house is built. It has only to be conveyed where it is wanted and used for growing.

271. **Some Forms and Changes of Vegetable Matter.** It may be used at once, or it may be stored up until it is wanted. In annual herbs, as already explained (68), nearly all of it is used for growth or for blossoming, as fast as it is made. In biennials, like the Beet, Carrot, and Turnip (70), a great part of it is stored up somewhere, generally in the root, and used the next year. In such perennials as the Potato, a part is laid up in the tubers (which are all of the plant that survives the winter), to begin a vigorous growth the next season. In shrubs and trees a part is annually deposited in the newest wood and bark, to be used for developing the buds the next spring. In all, a portion is deposited, as we know, sometimes in the fruit, always in the seed, for the use of the embryo or new plant, at the beginning of its growth.

272. When vegetable matter is laid up for future use, a large part of it is generally in the form of starch. Nearly the whole bulk of a potato, or of a grain of corn, is starch. This consists of little grains which are like mucilage solidified, and they may be turned into mucilage again. When the plant takes up a deposit of starch into its system, as fast as it dissolves it in the sap it generally changes it into sugar. Mucilage, starch, sugar, and plant-fabric, all have the same chemical composition, or very nearly ; and the plant readily changes one into the other as it needs. Notice the changes of vegetable matter in a plant of Indian Corn. In the leaves, where it is made, the elaborated sap is in the form of mucilage ; in the stalk, at flowering-time, while on its way to form and nourish the blossoms and grains, it turns sweet, being changed into sugar ; in the grain, a part is changed into starch and laid up there : when the grain germinates, the starch is dissolved and changed back into sugar ; and in the growing plantlet which it nourishes, the sugar is at length changed into plant-fabric.

273. **Circulation or Conveyance of Elaborated Sap, or Dissolved Vegetable Matter.** The new-made vegetable matter rarely accumulates in the leaves where it is made, except in the Century-plant, Houseleek (Fig. 65), and other fleshy-leaved plants. It is generally distributed through all the plant (that is, through all its living parts), or carried especially to where a stock is to be laid up, or where growth is taking place. So the elaborated sap, passing out of the leaves, is received into the inner bark, at least in trees and shrubs,—or in herbs it may descend through the soft parts generally,—and a part of what descends finds its way even to the ends of the roots, and is all along diffused laterally into the stem, where it meets and mingles with the ascending crude sap or raw material. So there is no separate

circulation of the two kinds of sap; and no crude sap exists separately in any part of the plant. Even in the root, where it enters, this mingles at once with some elaborated sap already there, and as it rises through the stem they mix more and more. But whatever is not assimilated must be, before the plant can use it; that which has been assimilated, can be used wherever it may be.

274. The elaborated sap, like the crude, is contained in the cavities, cells, or short closed tubes which make up the fabric of the plant, and circulates or passes from one to another through the partitions. How it passes through, and what attracts it where it is wanted, we do not very well know. And what we do know could not be well explained to the young beginner, for whom this book is written. The same may be said about

275. Growth. Growth is the increase of a living thing in size and substance. In plants it is done by the increase in the number of the *cells* (or cavities divided off by partitions) which make up the fabric, and by the increase of each in size to a certain extent. That is, growth is the building-up of the plant, or of additions to it, out of vegetable matter. And this vegetable matter was made in the leaves (either of the same plant or of its parent), out of mineral matter drawn from the earth and the air, — was mostly made of air and water. For the earthy part which is left behind when we burn a plant (and so turn all the vegetable matter back into air and vapor of water again) forms only a very small part of its bulk.

276. If the pupil would learn more particularly how growth takes place, and how plants change mineral into vegetable matter, they must study three or four lessons of the Lessons in Botany already referred to. But our short and simple account of the plant in action, i. e. *vegetating*, is sufficient for answering the main question, viz.: —

277. What Plants do. *Vegetation* consists essentially of two things, namely, *assimilation* and *growth*. In assimilation plants are changing mineral matter — air, water, and a little earth — into vegetable matter; and in growth this vegetable matter is wrought into all manner of beautiful and useful forms. This is the work which the vast variety and infinite number of plants over all the earth are busily engaged in. It is *their peculiar work;* for only plants can live upon (or assimilate) mineral matter; they only have the power of changing air, water, and earth into organic matter.

278. What is the effect of this action of plants, especially upon the air we breathe? And what becomes of all the vast amount of vegetable matter which

plants have been making, day by day, since God said, *Let the earth bring forth grass, and the herb yielding seed, and the fruit-tree yielding fruit after his kind, whose seed is in itself, upon the earth, — and it was so?* The answer to these questions will show us plainly

279. **What Plants are made for.** In the first place, in the very act of making vegetable matter, plants fulfil one great purpose of their existence, that is,

280. *They purify the air for animals.* That part of the air which renders it fit for breathing is called *oxygen ;* this makes up about one fifth part of the air we breathe. At every breath animals take in some of this *oxygen* and change it into *carbonic acid ;* that is, they combine the oxygen with carbon from their blood, which makes carbonic acid, and breathe out this carbonic acid into the air, in place of the oxygen they drew in. Now this carbonic acid is unfit for the breathing of animals, — so much so, that, if it were to increase so as to make any considerable part of the atmosphere, man and other animals could not live in it. But plants prevent the carbonic acid from accumulating in the air. While animals need the oxygen of the air, and in using it change it into carbonic acid, hurtful to them, plants need the carbon of this carbonic acid ; indeed, it makes a very large portion of their food, — as we plainly see it must, when we know that about half of every part of a plant is carbon, that is, charcoal. And this carbonic acid is the very part of the air that plants use; they constantly take it from the air, decompose it in their leaves during sunshine, keep the carbon, and give back the oxygen pure, so keeping the air fit for the breathing of animals. The carbon which plants take from the air in this way, along with water, &c., they assimilate, that is, change into vegetable matter : and in doing this

281. *They make all the food which animals live upon.* Animals cannot live upon air, water, or earth, nor are they able to change these into food which they may live upon. This work is done for them by plants. Vegetable matter in almost every form — especially as herbage, or more concentrated in the accumulations of nourishment which plants store up in roots, in bulbs and tubers, in many stalks, in fruits, and in seeds — is food for animals. " *And to every beast of the earth, and to every fowl of the air, and to everything that creepeth upon the earth,*" as well as to men, is given " *every green herb for meat.*" Some animals take it by feeding directly upon vegetables ; others, in feeding upon the flesh of herbivorous animals, receive what they have taken from plants. Man and a few other animals take in both ways what plants have prepared for them. But however received, and how-

ever changed in form in the progress from plant to animal or from one animal to another, all the food and all the substance of all animals were made by plants. And this is what plants are made for.

282. Notice also that plants furnish us not merely needful sustenance, but almost every comfort and convenience. *Medicine* for restoring, as well as food for supporting health and strength, mainly comes from plants.

283. *They furnish all the clothing of man ;* — not only what is made from the woolly hairs of certain seeds (*cotton*), or from the woody fibres of bark (*linen*), and what is spun from Mulberry-leaves by the grubs of certain moths (as *silk*), but also the skin and the fur or wool of animals, owe their origin to plants, just as their flesh does.

284. *They furnish utensils, tools, and building materials,* in great variety ; and even the materials which the mineral kingdom yields for man's service (such as iron) are unavailable without vegetables, to supply fuel for working and shaping them.

285. *They supply all the fuel in the world;* and this is one special service of that vegetable matter which, in the solid form of wood, does not naturally serve for food. Burned in our fire-places, one part of a plant may be used to cook the food furnished by another part, or to protect us against cold ; or burned under a steam-boiler it may grind our corn, or carry us swiftly from place to place. Even the coal dug from the bowels of the earth is vegetable matter, the remains of forests and herbage which flourished for ages before man existed, and long ago laid up for his present use. We may proceed one step farther, and explain where the heat of fuel comes from ; for even a child may understand it. Plants make vegetable matter only in the light, mostly in the direct light of the sun. With every particle of carbonic acid that is decomposed, and vegetable matter that is made, a portion of the sun's heat and light is absorbed and laid up in it. And whenever this vegetable matter is decomposed, as in burning it, this heat and light (how much of each, depends upon the mode of burning) are given out.

286. So all our *lighting* as well as warming, which we do not receive directly from the sun, we receive from plants, in which sunlight has been stored up for our use. And equally so, whether we burn olive-oil or pine-oil of the present day, or candles made from old peat, or coal-gas, or lard, tallow, or wax, — the latter a vegetable matter which has been somewhat changed by animals. And, finally,

287. *The natural warmth of the bodies of animals comes from the food they eat, and so is supplied by plants.* A healthy animal, no longer growing, receives into

his system a daily supply of food without any corresponding increase in weight, or often without any increase at all. This is because he decomposes as much as he receives. If a vegetable-feeder, far the greater part of his food (all the starch of grain and bread, the sugar, oil, &c.), after being added to the blood, is decomposed, and breathed out from the lungs in the form of carbonic acid and water. That is just what it would become if set on fire and burned, as when we burn oil or tallow in our lamps or candles, or wood in our fire-places; and in the process, in animals no less than in our lamps and fire-places, the heat which was absorbed from the sun, when the vegetable matter was produced from carbonic acid and water, is given out when this matter is decomposed into carbonic acid and water again. And this is what keeps up the natural heat of animals. We are warmed by plants in the food we consume, as well as by the fuel we burn.

288. In learning, as we have done, How Plants Grow, and Why they Grow, have we not learned more of the lesson of the text placed at the beginning of this book, and of the verses that follow? "*Wherefore, if God so clothe the grass of the field, shall he not much more clothe you? . . . . Therefore take no thought, saying, What shall we eat? or, What shall we drink? or, Wherewithal shall we be clothed? For your Heavenly Father knoweth that ye have need of all these things.*" And we now perceive that causing plants to grow is the very way in which He bountifully supplies these needs, and feeds, clothes, warms, and shelters the myriads of beings He has made, and especially *Man*, whom He made to have dominion over them all.

CHAPTER IV

Section I. — Classification.

289. CLASSIFICATION in Botany is the arrangement of plants according to their kinds and according to their resemblances.

290. In order that the vegetable creation might be adapted to every soil, situation, and climate, and to the different wants of the greatest variety of animals, as well as to the many peculiar needs of mankind, God created plants in a vast number of kinds. And in order that these should be perpetuated and kept distinct, He ordained that each should yield fruit and seed "*after its kind.*" So each sort of plant multiplies and perpetuates itself from generation to generation. Each of these sorts is a

291. **Species.** The individuals, or separate plants, of each sort represent that species, just as men and women represent the human species. The individuals of the same species are not always, or not commonly, exactly alike. They may differ in size according to their greater or less vigor ; they may vary in the color of their blossoms, or the shape and flavor of their fruit, and yet plainly be of one species. It is very apt to be so in cultivated plants. The different sorts of Apples belong to one species ; all the sorts of Pears are of one species ; and so of Peaches. Such sorts, which have arisen in the course of time and under change of circumstances, are called VARIETIES.

292. Varieties may be kept up with certainty by propagating from buds, that is, by cuttings, grafts, offsets, and the like (Chap. II. Sect. I.), but not by seeds, — at least when left to themselves. And varieties have nothing definite about them, but shade off into one another ; while the species are always separate. Apple-trees never vary into Pear-trees, nor Pear-trees into Quince-trees. The cultivator pays much attention to varieties, and takes particular pains to preserve and multiply them. To the botanist, who is concerned mainly with wild plants, they are of much less account. The botanist studies *species.*

293. According to their degrees of resemblance species form *Genera, Orders* or *Families,* and *Classes.*

294. **Genus:** plural **Genera.** Species which are very much alike belong to the same genus. The genus is a group of species which have the flower and fruit constructed on exactly the same plan. The Cabbage and the Turnip belong to the same genus. The different species of Raspberry and Blackberry belong to one genus, — the *Bramble genus*. The different species of Roses compose the *Rose genus ;* of Oaks, the *Oak genus ;* and so on.

295. **An Order or Family** (the two words meaning the same thing in Botany) is a kind of genus on a wider scale, consisting of genera, just as a genus consists of species. For example, while all the Oaks belong to the Oak genus, there are other trees which are a good deal like Oaks in the whole plan of their flowers, fruit, and seeds, so much so that we say they belong to the Oak family. Among them are the Chestnut, the Beech, and the Hazel ; each a genus by itself, containing several species. So the Pear genus, the Quince genus, the Hawthorn genus, the Rose genus, and the Bramble genus, with many more, belong to one great order. The Pea genus, the Bean genus, the Locust genus, the Clover genus, and the like, make up another order.

296. **A Class** is a great group of orders or families, all on the same general plan. The Rose family, the Oak family, and a hundred others, all belong to one great class. Lilies, Amaryllises, Irises, Palms, Rushes, and Grasses belong to another great class.

297. There are other divisions ; but these are the principal ones in all classifications, both of the vegetable and of the animal kingdom. And these four stages always rank in this way : the species under the genus, the genera under the order or family, and orders under the class, viz. : —

CLASS,

ORDER OR FAMILY,

GENUS,

SPECIES.

SECTION II. — Names.

298. THE name of any plant is the name of its genus and of its species. The name of the genus answers to the surname or family name of people, as *Smith* or *Brown.* The name of the species answers to the baptismal name, as *John* or *James.* We distinguish persons by these two names, as *John Smith* and *James Smith ; John*

Brown and *James Brown*, &c. In the same way, we name a plant by giving the name of the genus along with that of the species ; as *White Oak, Red Oak, Water Oak*. Here the first word is the name of the species, which is nothing by itself, but joined to the second word, which is the name of the genus, it designates the species of Oak ; and the two together completely name the plant we mean. These are

299. *Popular Names*, or the common names in our own language. Plants also have truly *Scientific Botanical Names*, which are the same in all countries. On this account they are in Latin. Some of them are the ancient Latin or Greek names ; others are words made in later times, but all are in Latin form. Thus, the scientific name of the Oak genus is *Quercus ;* of the Ash genus, *Fraxinus ;* of the Rose genus, *Rosa ;* of the Pear genus, *Pyrus ;* of the Bramble or Blackberry genus, *Rubus*, &c. The names of some genera are in honor of botanists or discoverers ; as, *Linnæa*, named in honor of Linnæus ; *Magnolia*, after Magnol ; *Kalmia*, after Kalm, a pupil of Linnæus, who travelled in this country ; *Claytonia*, after Clayton, a botanist of Virginia.

300. In the Latin or scientific name, that of the genus comes before the species. So the scientific name of the White Oak is *Quercus alba ;* of Red Oak, *Quercus rubra ;* of Water Oak, *Quercus aquatica*. In fact, these are just the popular names turned into Latin. It is not always so ; for what we call Post Oak is botanically named *Quercus obtusiloba*, which means an Oak with blunt lobes to the leaves. And our White Ash is *Fraxinus Americana*, meaning " American Ash "; Red Ash is *Fraxinus pubescens*, meaning " Downy Ash "; Black Ash is *Fraxinus sambucifolia*, meaning " Elder-leaved Ash." But our Green Ash is *Fraxinus viridis*, which means the same thing as the common name.

301. The name of the genus is a substantive. That of the species is generally an adjective ; as, *viridis*, green ; *sambucifolia*, Elder-leaved ; *Americana*, American ; *aquatica*, growing in water ; and so forth.

302. Accordingly, any plant is named in two words, that is, by giving the name of its genus and of its species.

303. The names of the class, order, &c. make no part of the name of the plant itself. And these names differ in different systems of classification, while those of the genus and species are the same in all systems.

7

SECTION III. — The Natural System.

304. There are two kinds of classification in Botany, viz. *Natural* and *Artificial.* They differ in the way the genera are arranged in orders, classes, &c.

305. **An Artificial Classification** is one in which plants are arranged for convenience of reference, or for finding out their names, without any particular care for bringing like things together. Tournefort made an artificial classification of plants by their flowers, mainly by their corolla, which was in common use in the last century until Linnæus contrived a better one, in which the classes and orders were founded upon the number, position, &c. of the stamens and pistils. This was in general use for many years. But now we use artificial classifications only in the form of *Tables* or *Analyses*, as a key for finding out the family a plant we are studying belongs to, and so readily referring it to its place in

306. **The Natural System.** In this system plants are classified according to their relationships, that is, according to their resemblances in all respects. The most important resemblances are used for the classes, &c.; the most important after these for the orders; more particular ones mark the genera; and matters of shape, proportion, color, &c. mark the species. So the whole together forms a *system*, in which all known plants are to be ranked in their natural order, each standing next those which it is most like in all respects; the whole forming, as it were, a great map, in which the classes and other great divisions might answer to countries, the orders to counties, and the genera to towns or parishes.

307. Such a system is not a mere convenience for ascertaining the name of a plant, but is an illustration, as far as may be, of the *plan of the Creator* in the vegetable kingdom. And the Botanist sees as much to admire, and as plain evidences of design, in the various relations of the species of plants to each other (i. e. in their resemblances and their differences), as he does in the adaptation of one part of a plant to another, and in the various forms under which any one organ may appear. The different kinds of plants are parts of a great whole, like the members of a body, or the pieces of an harmonious but complex edifice or structure; and this whole is the *Vegetable Kingdom.*

308. What the main divisions in the system are, may be gathered from what is stated in several places in Part I. In the first place, the whole vegetable kingdom divides into two great *Series* or *Grades*, — a higher and a lower. The higher series contains all

-FLOWERING OR PHÆNOGAMOUS PLANTS, namely, those that are propagated by means of real flowers, producing seeds, which contain an embryo ready formed. The lower series consists of

FLOWERLESS OR CRYPTOGAMOUS PLANTS, which produce no real flowers and no true seeds, but only something of a simpler sort, answering to flowers and giving rise to *spores*, which serve the purpose of seeds.

309. This has been explained in Chapter II. Section II. p. 58. Next, the great series of Flowering Plants is divided into two *Classes*. These classes are distinguishable by the stem, the leaves, the flower, and the embryo or germ of the seed. They are : —

Class I. EXOGENS, or DICOTYLEDONS (more fully named, *Exogenous or Dicotyledonous Plants*). Plants of this class, as to their *stems*, have the wood all between a separate pith in the centre and a bark on the surface, and each year the stem lives, it forms a new layer of wood on the surface of that of the previous year (111, 115 – 118). As to the *leaves*, they are netted-veined or reticulated, the veins branching and forming meshes (126, 127). As to the *flowers*, their parts are generally in fives or fours (or the double or treble of these numbers), very rarely in threes. As to the *embryo*, or germ, it always has a pair of cotyledons or seed-leaves (48), or sometimes more than a pair (49).

Class II. ENDOGENS, or MONOCOTYLEDONS (or more fully, *Endogenous* or *Monocotyledonous Plants*). Plants of this class, as to their *stems*, have their wood in threads mixed with the pith and scattered throughout every part, never forming layers, and the bark is never to be peeled off clean from the wood (112 – 114). The *leaves* are almost always parallel-veined (127 – 129). The *flowers* have their parts in threes (or twice three), very rarely in twos or fours, never in fives, which is much the commonest number in the other class. And the *embryo* has but one cotyledon or seed-leaf (47, 50).

310. So the class of any plant may be told from a piece of its stem alone; or from a single leaf, in most cases; or from a blossom; or from a seed; or from the plantlet as it springs from the seed, and in its first leaves shows the nature of the embryo. The seeds generally are not easy to study without a dissecting microscope, nor can we always have them growing. But the student will hardly ever fail to tell the class at once, by the stem, the leaves, or the flowers, and by the whole look of the plant.

311. The first Class divides into two *Subclasses*, of very unequal size, viz.: —

Subclass I. ANGIOSPERMS (or *Angiospermous Plants*), which have pistils of the common sort, in which the seed is formed and contained (16, 219). This takes all of the first class except the Pine family, and one or two small orders little known in this country. These form the

Subclass II. GYMNOSPERMS, that is, *Gymnospermous* or *Naked-seeded Plants* (218, 250). Here the ovules and seeds are naked, there being no pistil at all, as in the Yew, or only an open scale that answers to it, as in Pines, Cedars, &c.

312. The first class contains about a hundred common orders or natural families; the second not half so many.

313. The lower or second series, that of *Flowerless* or *Cryptogamous Plants*, divides into three classes, viz.: —

Class III. ACROGENS, which includes the Fern family, the Horsetail family, and the Club-moss family.

Class IV. ANOPHYTES, which consists of the orders of Mosses and Liverworts.

Class V. THALLOPHYTES, which includes Lichens, the Algæ or Seaweeds, and the Fungi or Mushroom family.

314. But Flowerless plants, being too difficult for the beginner, need not be further mentioned here.

315. The orders or families in the natural system are pretty numerous. They are named, in general, after some well-known genus which may be said to represent the family. Thus the order to which the Rose belongs is called the *Rose family*; that to which Crowfoots or Buttercups belong, the *Crowfoot family*; that to which Cress and Mustard belong, the *Cress family*; the Oak gives its name to the *Oak family*, the Birch to the *Birch family*, the Pine to the *Pine family*, and so on. Their Latin or scientific names are also generally made from the Latin name of a leading or well-known genus. For example, *Rosa*, the Rose, gives its name to the Rose family, viz. *Rosaceæ*, meaning Rosaceous plants; *Ranunculus*, the Crowfoot genus, gives to its family the name of *Ranunculaceæ*; and *Papaver*, the Poppy, gives to its family that of *Papaveraceæ*; *Berberis*, the Barberry, that of *Berberidaceæ*; and so on.

316. The student's principal difficulty at the beginning will be to find out the order or family to which a plant belongs. This is because the orders are so numerous, and commonly not to be certainly distinguished by any one point. But after some practice, the order will be as easy to make out as the class; and in many cases it will be known at a glance by the strong family likeness to some plant which has been examined before.

317. Let us now introduce our pupils to the *Popular Flora*, by which they may study the common plants they meet with, and find out their structure and their names.

Section IV. — How to study Plants by the Popular Flora.

318. **Directions for gathering Specimens to Examine.** Gather specimens with flowers, flower-buds, and also with fruits, either forming or full grown, when all these are to be had at the same time, as they frequently are, at least in herbs, except in spring. Sometimes the remains of last year's fruit are to be found, enough to tell what the kind of fruit was. Very often the nature of the fruit can be told beforehand, from the pistil, either at flowering-time or soon after. However, most of our common plants may be made out from the blossoms and leaves only. Small herbs should be taken up by the roots.

319. Specimens which are to be kept for some time, or carried to some distance before they are studied, should be put at once into a close-shutting tin box, where they will keep long without withering. Botanical boxes are made for the purpose. A candle-box, or any tin box with a lid, and of convenient size for carrying, will answer.

320. **For examining Plants** to make out the structure of the flowers, fruits, &c., the instruments most needed are, —

A sharp, thin-bladed pocket-knife, such as a common penknife, for making sections or slices;

A pair of small forceps, which, although not always necessary, are very convenient for holding little parts; and

A hand microscope or magnifying-glass, such as may be purchased for a dollar or less. A single glass, mounted in horn, or in metal, and (for carrying in the pocket) shutting into a case of the same material, which serves as a handle when open, is the commonest and best for our purpose.

A stand-microscope is a most convenient thing, when it can be had. This has a glass stage under the lens or magnifying-glass, on which small flowers, or their parts, may be laid. This leaves both hands free for dissecting or displaying the minute parts, with needles mounted in handles, while the eye is examining them under the microscope. Common needles, mounted in the bone handles used for holding *crochet* needles, are very convenient, and cost little. A compound microscope, however necessary for studying vegetable anatomy, is of no use for our common botanical purposes, which require no high magnifying. A pocket magnifying-glass, held in the hand, is all that is absolutely necessary.

321. **Lessons in examining Plants.** How a pupil, or a class, is to proceed in examining any plant by our *Popular Flora*, for the purpose of finding out its class, its order or family, and then its genus and species, — that is, its name, — we will show by a few plain examples.

322. Technical words or terms are used all along, which you may not remember the meaning of, as defined in the first part of the book ; and some of them may not have been mentioned or explained there. Whenever you come to a word which you do not *perfectly understand*, turn at once to the Index and Dictionary, beginning on page 217, and look it out. There you will find it explained, or will be referred to the page of the book where the term is explained or illustrated. Turn back to the place, and read what is said about it. Do not attempt to proceed faster than you understand things. But by looking out and understanding the words as you meet with them, the principal terms used in botanical descriptions (here made as simple as possible) will soon be familiar, and your subsequent progress will be all the more rapid for the pains taken in the earlier steps.

323. For the first example we will take a Buttercup or Crowfoot, such as may anywhere be met with in spring and early summer. With specimens in hand, turn to page 105. You ask in the first place, —

Does the plant belong to the *First Series*, that of Phænogamous or Flowering Plants? Certainly ; for it bears flowers, with stamens and pistils. (The Second Series, that of Flowerless Plants (p. 97), consisting of Ferns, Mosses, &c., we do not meddle with in this book, they requiring too much magnifying, and being too difficult for the young beginner.) Next you ask, —

To which class does it belong? The differences between the two classes are mentioned on page 97, and the characteristics of Class I. are illustrated on page 105. As the stem is hollow, it may not be easy to see that it has a delicate ring of wood under the bark and outside of the pith (as in Fig. 230) ; but this may be perceived in a cross slice under the microscope. And even if we had ripe seeds, a microscope and some skill in dissection would be required to take out the minute embryo, and see that it has a pair of cotyledons. But we may tell the class by the two other points, viz. by the leaves, and by the number of parts to the blossom. The leaves are plainly netted-veined, and the parts of the flower, that is, the sepals

and the petals, are five. So the plant belongs to Class I. Proceed then with the
"Key to the Families or Orders of Class I." on page 106. This class has two
subclasses. So you next ask, —

To which subclass does the plant belong, to Angiosperms or to Gymnosperms?
For the character of the Gymnosperms, see the end of the Key, at the foot of
page 111; that of Angiosperms begins the Key. The centre of the flower we are
examining is occupied with a great number of small one-seeded pistils, each tipped
with its short style and stigma; and the ovary is a closed bag containing an ovule
or young seed. So the plant clearly belongs to Subclass I. Proceed then with
the Key; which leads you next to ask, —

To which division does the plant belong, — the *Polypetalous?* (in black letters
immediately under the subclass), or the *Monopetalous?* (top of page 109), or the
Apetalous? (lower part of page 110). Plainly to the first or *Polypetalous* division;
for there is both a calyx and a corolla, and the latter is of five separate petals.

This division, in the Key, subdivides into, "A. Stamens more than 10," and
"B. Stamens 10 or fewer" (p. 107). Our plant has many stamens, and so falls
under the head A.

This head subdivides into three (marked 1, 2, 3), by differences as to where and
how the stamens are borne. Pull off the calyx and the corolla, or split a flower
through the middle lengthwise (as in Fig. 238), and you will plainly see that the
stamens stand on the receptacle, under the pistils, unconnected either with the calyx
or the corolla. So the plant falls under the head 1.

Under this is an analysis of some of the characters (i. e. distinguishing marks) of
the fifteen or sixteen families which belong here. The lines that are *set in* are
subdivisions under the longer line above them. The lines which rank directly under one another (and begin with the same or a corresponding word) make alter-
natives, among which you are to choose that with which your plant agrees. In
this instance the lines of the first rank here begin with the word "Pistils" or
"Pistil," and there are five of them. Try the first: "Pistils more than one,
entirely separate from each other." That is the case with our plant. Under this
line, in the next rank, is a triplet, or a choice between three. Our plant is an "herb,
with perfect flowers," and so falls under the first line. Under this is a couple
of equivalent lines, relating to the leaves. Our plant agrees not with the second,
but with the first of these; and that line ends with the English name of the
family we are seeking for, viz. the CROWFOOT FAMILY, and refers to page 112,
where this family is described.

Turn now to the account of this family, and read over the descriptive marks given, to see if you have been led to a right conclusion. The description agrees, as far as it goes. Knowing the family, you now ask,—

To what genus of this family does the plant belong? The genus gives the principal name of the plant; so this is the same as asking, What is the plant's name? Now, in every family which has several genera or kinds under it, we have a key to the genera, like that which we have just used under the class to find out the family. Try the key, then, under this family, to find out the genus.

This key begins with a pair of lines, viz. "Climbing plants," &c., and "Not climbing," &c. Our plant agrees with the latter. Under this, in the next rank, is a pair of lines, beginning with "Pistils" (the second line of the pair is the sixth on p. 113). You perceive that our plant falls under the first. Under this is the line beginning "Petals none." Our flower has petals; so pass on to the other one of the pair, which is the fifth line on p. 113. This reads "Petals present as well as sepals, the latter falling off early" (which agrees); and leads to the name of the genus, i. e. "(*Ranunculus*) CROWFOOT."

The first name, in parentheses and in Italic type, is the scientific or Latin name of the genus; the other, in small capitals, is the popular English name of the genus. When we have only one species to the genus, we do not in this book proceed farther. But there are many Crowfoots, so you next inquire,—

What is the species? Look on, till you come to the name of the genus in dark letters, on p. 114. Here a few more marks of the Crowfoot genus are given; and then the marks of ten common species of Crowfoot follow, under several heads. We are supposed to have in our hands one of the two large yellow-flowered species, commonly called Buttercups. Compare the specimens with the divisions marked by stars. It cannot belong to that with one star, for the petals are not white; it does belong to that with two stars, for the petals are yellow, and bear a little scale on the inside just above the bottom. Under this are two divisions, marked with daggers. Not growing under water, our plant belongs to that marked + +. Under this are two further divisions, marked ++ and ++ ++ : our plant, having the "petals much longer than the calyx," belongs to the second of these.

Under this head are four species. The English name is given at the beginning of the line, in small capital letters; a short description follows, and the scientific or Latin name is appended, in Italic letters, at the end. Here the *R.* of course stands for *Ranunculus*. A comparison with the description will show which species it is that we happen to have. If a field plant flowering in May, and with a bulbous base of the stem just underground, it is the BULBOUS CROWFOOT or BUTTERCUP, or in Latin, *R. bulbosus*. If the taller species, without a bulb, and flowering in summer (which is the most common kind throughout the country), it is TALL CROWFOOT or BUTTERCUP, or *R. acris*. Having in this way made out one Crowfoot, you will be sure to know any other one as soon as you see it, and will only have to find out the species, comparing your specimen with the descriptions, on p. 114.

324. Suppose, for the next example, you have specimens, with flowers and young fruit, of a common plant in wet grounds in spring, here called Cowslip, though this is not its correct English name. With specimens in hand, turn to p. 105.

To which class does it belong? Its netted-veined leaves (and the structure of the stem, as seen in a slice under a good magnifying-glass) plainly refer it to Class I. You next ask, —

To which subclass? The pistils and pods plainly refer it to Subclass I.

To which division? At first view you may think it has a corolla; but there is no calyx outside of these yellow leaves of the flower, even in the bud. So you will conclude that these leaves are the calyx, notwithstanding their rich color and petal-like appearance; and you will turn to the *Apetalous* division, on p. 110.

Continue the analysis under that division. The flowers are separate, and "not in catkins"; so it falls under A. The seeds are numerous in each ovary or pod; so it falls under No. 1. The "calyx is free from the ovary," according to the second of the first pair of lines. So you have only to choose between the three lines of the triplet under this, beginning with "Pod." As the pistils and pods are one-celled and simple, we are brought to the name †CROWFOOT FAMILY, p. 112. The mark † denotes that you have in this case an apetalous plant belonging to a family in which the flowers generally have petals. You turn to this family, p. 112, and proceed as before. You are led along the same track, until you reach the line "Pistils many or several, becoming akenes in fruit." Your flowers have a number of pistils, but these contain numerous seeds, and make pods in fruit, as in Fig. 240. So you pass on to the other line of the couplet, which reads, "Pistils more than one-seeded, becoming pods"; which agrees with the plant in hand. The first line in the next rank reads: "Sepals petal-like, not falling when the flower first opens" (so it is in your plant); and, of the four lines of the next rank, you can take only the first: "(Sepals) golden-yellow: petals none: leaves rounded, not cut." This brings you to the name of the genus, — in Latin or scientific form, *Caltha;* in English, MARSH-MARIGOLD. Being the only species, we need go no farther with it.

325. On reflection and comparison, you will perceive the family likeness between the Marsh-Marigold and the Crowfoot, different as they are in some particulars; and between these and the Globe-flower, the Gold-thread, the Anemony, and even the Larkspur and Aconite, when you have studied these plants. But the family likeness is not quite so strong at first view in this family as it is in most others.

326. Another example we will take from the plant figured on p. 5 and the following pages (Fig. 4 – 19), a very common ornamental twiner about houses, flowering all through the summer. Begin, as before, on p. 103. You perceive at once that the plant belongs to Class I.; for it has netted-veined leaves, the parts of the flower are in fives, and the embryo (which is easily extracted from the fresh seed, Fig. 16 – 19) has a pair of seed-leaves. There is a regular pistil, and the seeds in a pod; so the plant belongs to Subclass I. There is both calyx and corolla, the latter of one piece; so the plant belongs to the Monopetalous division, p. 109. The corolla is borne on the receptacle below the ovary; so you pass to the head B. The stamens are just as many as the lobes, or rather here the plaits, of the corolla; so you pass No. 1, and take No. 2. The stamens stand before the plaits, so that they would be alternate with the divisions of the corolla, if it were not that the five petals it consists of are united to the very top; so you take the second of the two lines commencing with the word "Stamens." These are "inserted on the corolla," and are entirely separate and "free from the stigma"; so you take the fourth line of those in the next rank. There is a style (p. 110); so the plant falls under the second of the two lines of the next rank. The ovary and pod have 3 cells; so it falls under the third of the lines beginning with the word "Ovary." The stamens are 5, and the pod few-seeded (2 seeds in each cell); so it falls under the third of the lines beginning with "Stamens." The plant twines, and the seeds are large; so you are brought to the name of the family, the CONVOLVULUS FAMILY, and are referred to p. 184. Read over the marks of the family, and then search for the genus in the key or arrangement; and you will find that the name of the genus is, in scientific language, *Ipomœa,* in popular English, MORNING-GLORY.

327. One more example, to show how plants are to be studied by the Flora, will be sufficient. Take the Lily of the Valley (Fig. 3 on p. 1), which in this country adorns almost every flower-garden.

328. With plants in hand, turn to p. 105, and compare with the distinguishing marks of Class I. A slice across the stem shows no ring of wood around a pith. The leaves are not netted-veined. The parts of the flower are not in fives or fours, but in sixes, that is, twice threes. So the plant does not agree with Class I. in any respect. Turn therefore to Class II., on p. 203. Examining slices of the stem with a magnifying-glass, you may find threads of wood interspersed in the cellular part or pith. The leaves are parallel-veined (Fig. 502, 503). The flowers have their parts in threes or twice threes; i. e. the cup of the blossom has six lobes, and there are six stamens; and, although there is only one pistil, the stigma is three-lobed and the ovary has three cells, showing that it is composed of three pistils grown into one. So, without looking for the embryo in a ripe seed, which is not often to be had, you are sure the plant belongs to Class II. ENDOGENS or MONOCOTYLEDONS.

329. To find out the family or order the plant belongs to, try the Key. There are three divisions of the class. First, the *Spadiceous*, which has the flowers sessile on a spadix or fleshy axis. Not so with the plant in hand, which has drooping blossoms in a slender raceme. Pass on, therefore, to the second or *Petaloideous* division. In this the flowers are not on a spadix, nor enclosed in chaffy bracts or glumes, and they have a calyx and corolla, or a perianth colored like a corolla. Our plant belongs to this division. The first line under it reads: "Perianth free from the ovary"; this is the case in our plant. Proceed to the next rank: "Of 3 green or greenish sepals and 3 distinct and colored petals." Not so in our plant; so we pass to the corresponding line: "Of 6 petal-like leaves in two ranks, or 6-lobed and all colored alike." Here our plant belongs. Proceed to the two lines under this, beginning with the word "Stamens." Our flowers have six stamens; so we take the second line of the pair. Pass to the two lines of the next rank, beginning with "Anthers." These in our plant are turned inwards: so we take the second line of the pair, and are led to the Lily Family, p. 209. Turn to that page: read over the marks of the family, and go on to ascertain the genus. Having few seeds or ovules in the ovary, small flowers, and running rootstocks, we find our plant to agree with the first line of the key to the genera of the Lily Family. The simple and naked scape or flower-stalk from the ground, &c. accords with the third line of the next rank; and the flowers in a raceme answer to the first of the two lines under that. And this brings us to the name of the genus, viz. in Latin form, *Convallaria*; in English, LILY OF THE VALLEY, — the only species of the genus.

330. Signs and Abbreviations used in the Popular Flora. These are very few and easily understood.

The signs for degrees (°), minutes ('), and seconds (") are used for size or height; the first for feet, the second for inches, and the third for lines or twelfths of an inch.

Accordingly 1° or 2° means one or two feet long or high, as the case may be.

And 1' or 2' means one or two inches long or high.

And 1" or 2" means one or two lines or twelfths of an inch long.

An asterisk or star before the name of a genus — as * FENNEL-FLOWER and * PEONY on p. 113, or * RADISH, * TURNIP, * CANDYTUFT, &c. on p. 125 — denotes that there are no wild species of that genus in this country, but they are to be met with only as cultivated plants.

§ This mark stands for section of a genus, or a *subgenus*, i. e. a section almost distinct enough for a genus. See under Magnolia, p. 117; also p. 147, where *Pyrus*, § *Sorbus*, and *Pyrus*, § *Malus*, &c. denote that *Sorbus* and *Malus* are only sections or subgenera of the genus *Pyrus*.

To save room, the name of the genus generally is not printed in full under each species. So, under Virgin's Bower, p. 113, the first species, WILD VIRGIN'S BOWER, is given in full. In the second, "SWEET V." stands for Sweet Virgin's Bower. Also, as to the scientific name, "*C. Flammula*" stands for *Clematis Flammula*, — and so elsewhere.

N., S., E., and W., which are occasionally added after the description of a species, stand for North, South, East, and West, and indicate the part of the country where the plant naturally grows. For example, the LONG-FRUITED ANEMONY, p. 114, is found North and West (N. and W.), &c. When there is no such reference, the species may be found in almost any part of the Northern United States.

Fl. is an abbreviation for flowering, or sometimes for flower. P. 115, line 1, &c. "Fl. spring," means flowering in spring, "Fl. summer," line 8, means flowering in summer. Cult. is an abbreviation for cultivated.

Accents. In the Latin or scientific names, the syllable upon which the accent falls is marked with a ' or `. When the accented vowel has a long sound, it is marked `; as *Anemòne*, p. 115, *Aconìtum*, p. 116. When the vowel has the short sound, it is marked '; as *Clématis* and *Hepática*, p. 115.

All Latin or Latinized names, when of only two syllables, take the accent on the first syllable, and therefore do not need to be marked.

Part Second.

POPULAR FLORA,

A CLASSIFICATION AND DESCRIPTION OF

THE COMMON PLANTS OF THE COUNTRY,

BOTH WILD AND CULTIVATED, UNDER THEIR
NATURAL ORDERS.

A FLORA is a botanical account of the plants of a country or district, with the orders or families systematically arranged under the classes, the genera under the orders, and the species (when there are more than one) under the genus they belong to, — along with the *characters* of each class, order, genus, &c.; that is, an enumeration of the principal and surest marks, or some of them, by which they are to be distinguished. A full Flora of all the plants which grow in this country, including those in common cultivation, would at the least fill a large volume; and would be both too expensive and too unwieldy for the young beginner. The *Manual of the Botany of the Northern United States* (including Virginia and Kentucky, and extending west to the Mississippi River) is a volume of over 600 pages, or 700, including the Mosses. And this work does not include foreign plants cultivated in our fields or gardens, except those that have run wild in some places.

The POPULAR FLORA, which occupies the rest of this book, is for the use of beginners, and is made as brief, simple, and easy as possible. For greater facility in the study, it includes only the common wild plants of the country (especially of the Northern States), and those ordinarily cultivated in our fields or gardens, for use or ornament. The families or genera which are too difficult for young beginners, such as Grasses, Sedges, the large family of plants with compound flowers (the Sunflower Family), and the like, are altogether omitted or only briefly

alluded to.   So also are the Cryptogamous or Flowerless Plants, as already
mentioned.   To save room, when there is only one species, or only one common
species, to a genus, we do not proceed any farther with it than to the name of the
genus, both scientific and popular.

Under the species the English or popular name is placed foremost, in small capitals ; the scientific or Latin name at the end.   The scientific names throughout are
printed in italic letters.

Full instructions for using the Flora in studying plants are given in Chapter
IV. Section IV.; at the close of which, the few abbreviations and signs employed
are explained.

Classes and other great Divisions.

POPULAR FLORA.

SERIES I.

FLOWERING OR PHÆNOGAMOUS PLANTS.

PLANTS which produce real Flowers (or Stamens and Pistils) and Seeds. — See Part I. Paragr. 164, 166.

CLASS I. — EXOGENS OR DICOTYLEDONS.

Stem composed of pith in the centre, a separate bark on the surface, and the wood between the two, of as many rings or layers as the stem is years old.

Leaves netted-veined, that is, with some of the veins or veinlets running together so as to form meshes of net-work or reticulations.

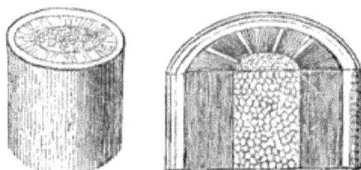

Exogenous stem of the first year.

. *Flowers* with their parts most commonly in fives or fours, very seldom in threes.

Embryo dicotyledonous, i. e. of a pair of seed-

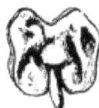  

232. Netted-veined leaves of Maple.   Embryos of, 233. Sugar-Maple ;   234, 235. Morning-Glory ;   236. Cherry.

leaves, or in the Pines and the like often polycotyledonous, that is, of more than one pair. — The class may be told by the stems and leaves without examining the seeds; but embryos are represented in the figures, to show the student what is meant. — For the other class, see p. 203.

KEY TO THE FAMILIES OR ORDERS OF CLASS I.

SUBCLASS I. — ANGIOSPERMS.

With a regular pistil, and a seed-vessel in which the seeds are formed. See Paragr. 219, 311.

I. Polypetalous Division. Calyx and corolla both present; the petals entirely separate.

A. STAMENS MORE THAN 10.

1. *Stamens on the receptacle, unconnected either with the calyx, corolla, or ovary.*

Pistils more than one, entirely separate from each other.
  Herbs, with perfect flowers. Page
    Leaves not shield-shaped, generally cut, toothed, or compound, CROWFOOT FAMILY, 112
    Leaves shield-shaped, fixed by their middle, WATER-SHIELD, 121
  Woody vines, with diœcious flowers, shield-shaped leaves fixed near the edge, MOONSEED F. 118
  Small trees with perfect flowers, 6 petals, and entire leaves, CUSTARD-APPLE F. 118
Pistils many, grown together one above or over another on a long receptacle, MAGNOLIA F. 117
Pistils several, sunk in the flat top of a broad receptacle, NELUMBO, 121
Pistils 3 to 6, the ovaries partly grown together in a circle, making
  A bladdery pod of several cells, Fennel-flower in CROWFOOT F. 112
  A several-horned one-celled pod, MIGNONETTE F. 125
Pistil only one, at least having only one ovary; and that
  Simple and one-celled, only one placenta or seed-bearing part.
    Petals 6 to 9, large. Leaves 1 or 2, many-lobed, May-Apple in BARBERRY F. 119
    Petals 4 and irregular, or else very small, CROWFOOT F. 112
  Compound, with many seeds on a placenta from the bottom of the cell, PURSLANE F. 130
  Compound, if one-celled, then with two or more seed-bearing lines on the walls.
    Calyx falling when the flower opens; sepals fewer than the petals, . POPPY F. 122
    Calyx falling after blossoming. Style 1: ovary several-celled.
      The 5 sepals edge to edge in the bud. Fruit dry, 1-seeded, LINDEN F. 133
      The sepals overlapping in the bud. Fruit many-celled, ORANGE F. 134
    Calyx remaining beneath the fruit.
      Leaves with transparent or dark colored dots, all opposite, ST. JOHN'S-WORT F. 128
      Leaves not dotted. Ovary and pod one-celled, CISTUS F. 127
      Leaves not dotted. Ovary several-celled. Aquatic or bog plants.
        Leaves pitcher-shaped. Style umbrella-like, SIDESADDLE-FLOWER F. 121
        Leaves rounded and heart-shaped. Style none, WATER-LILY F. 120

2. *Stamens connected with the bottom of the petals, and these borne on the receptacle.*

Filaments united in a pretty long tube or column: anthers kidney-shaped, one-celled, MALLOW F. 131
Filaments united only with the base of the petals: anthers oblong, two-celled, CAMELLIA F. 132

3. *Stamens borne on the calyx, or where the calyx (when coherent) separates from the ovary.*

Petals many, in several rows.
    Shrubs with opposite simple leaves and dingy-purple flowers,   Carolina-Allspice F. 152
    Leafless fleshy plants, of singular shapes,   Cactus F. 153
    Water-plants, with the large flowers and leaves floating on the surface,   Water-Lily F. 120
Petals 4 or 5, rarely 6.
    Leaves with stipules, alternate,   Rose F. 146
    Leaves without stipules. Pods many-seeded.
        Style and stigma one. Pod surrounded by the free calyx,   Lythrum F. 152
        Styles or stigmas 3 to 8. Calyx coherent below with the ovary.
            Shrubs: leaves opposite. Pod with several cells.   Philadelphus in Saxifrage F. 157
            Herbs: leaves fleshy. Pod one-celled, opening by a lid,   Purslane F. 130

B. Stamens 10 or fewer.

1. *Corolla irregular.* (Pistil one.)

Leaves opposite, palmately compound. Calyx 5-toothed. Shrubs or trees,   Horsechestnut F. 130
Leaves alternate, with stipules.
    Filaments often united, but not the anthers. Two lower petals approaching or joined.
                  Pod simple, with only one row of seeds,   Pulse F. 141
    Filaments short: anthers 5, united. Lower petal with a sac or spur at the base. Pod
                with 3 rows of seeds on the walls,   Violet F. 126
Leaves alternate, without stipules. Flower generally 1-spurred or 2-spurred.
    Stamens 5, short; their anthers a little united. Pod bursting at the touch,   Balsam F. 136
    Stamens 8, separate. Fruit of 3 thick and closed pieces,   Indian-Cress F. 135
    Stamens 6, in two sets. Flower closed. Pod one-celled,   Fumitory F. 123

2. *Corolla regular, or nearly so.*

Stamens just as many as the petals, and standing one before each of them.
    Pistils more than one, and separate. Petals 6. Flowers diœcious,   Moonseed F. 118
    Pistil with one ovary but with five separate styles,   Leadwort F. 173
    Pistil and style one (the latter sometimes cleft at the summit).
        Anthers opening by uplifted valves or doors. Petals 6 or 8,   Barberry F. 119
        Anthers not opening by valves, but lengthwise.
            Woody vines. Calyx minute: petals falling very early,   Grape-Vine F. 137
            Shrubs. Calyx larger, its divisions 4 or 5,   Buckthorn F. 138
            Herbs. Ovary and pod one-celled.
                Sepals 2: petals 5: stigmas 3,   Purslane F. 130
                Sepals as many as the petals: style single: stigma one,   Primrose F. 173
Stamens as many as the petals and alternate with them, or twice as many, or of some unequal number.
    Calyx with its tube adherent to the surface of the ovary.
        Stamens 3, united with each other more or less. Flowers monœcious,   Gourd F. 154
        Stamens distinct, as many or twice as many as the petals.

8

Seeds many in a one-celled berry. Shrubs,   CURRANT F. 155
Seeds many in a 2-celled or 1-celled pod: styles 2,   SAXIFRAGE F. 157
Seeds many: pod 4-celled: style 1: stigmas 4,   EVENING-PRIMROSE F. 153
Seeds (1 to 5) one in each cell. Border of the calyx obscure.
  Flowers in cymes or heads. Style and stigma one,   CORNEL F. 160
  Flowers in umbels.
    Umbels compound: styles 2: fruit dry,   PARSLEY F. 158
    Umbels simple or panicled: styles 3 to 5, rarely 2: fruit a berry, ARALIA F. 159
Calyx free from the ovary, at least from the fruit.
  Leaves punctured with transparent dots, sharp-tasted or aromatic.
    Leaves simple, all opposite and entire,   ST. JOHN'S-WORT F. 128
    Leaves compound,   RUE F. 137
  Leaves without transparent dots.
    Pistils more than one. Leaves with stipules,   ROSE F. 146
    Pistils 4 or 5. Herbs without stipules,   STONECROP F. 156
    Pistils 2, nearly distinct. Stipules none,   SAXIFRAGE F. 157
    Pistil one, simple, one-celled: style and stigma one,   PULSE F. 141
    Pistil one, compound, either its styles, stigmas, or cells more than one.
      Style one (in Cress F. often short or none), entire, or barely cleft at the top.
        Anthers opening by holes or chinks at the top, }
        Anthers opening across the top,   }   HEATH F. 168
        Anthers opening lengthwise.
          Herbs: stamens on the persistent calyx,   LYTHRUM F. 152
          Herbs: stamens on the receptacle, 6, two of them shorter,   CRESS F. 124
          Woody plants. Fruit few-seeded.
            Stamens fewer than the 4 long petals,   FRINGE-TREE, 189
            Stamens as many as the broad petals,   STAFF-TREE F. 139
      Styles or sessile stigmas 2 to 6, or style 2- to 5-cleft.
        Ovary and fruit one-celled, and
          One-seeded. Shrubs,   SUMACH F. 137
          Six-seeded on 3 projections from the walls,   PINWEED, 127
          Several- or many-seeded. Stamens distinct.
            Seeds in the centre of the pod. Leaves all opposite,   PINK F. 129
            Seeds on the walls or bottom of the pod,   SAXIFRAGE F. 157
          Many-seeded along the walls of a long-stalked berry.
            Stamens monadelphous,   PASSIONFLOWER F. 155
        Ovary with 2 to 5 or more cells.
          Sessile stigmas and stamens 4 to 6,   HOLLY F. 171
          Styles 3. Leaves opposite, compound,   BLADDERNUT F. 189
          Styles or long stigmas 2. Fruit 2-winged,   MAPLE F. 138
          Styles or divisions of the style 5.
            Stamens 5: pod partly or completely 10-celled,   FLAX F. 134
            Stamens 10: pod 5-celled. Leaves compound,   WOOD-SORREL F. 136
            Stamens 10 (or fewer with anthers): styles united
              with a long beak, splitting from it with the
              5 one-seeded little pods when ripe,   GERANIUM F. 135

II. Monopetalous Division. Corolla with the petals more or less united into one piece. (Those which rank in other divisions are marked †.)

A. COROLLA ON THE OVARY, i. e. tube of calyx coherent.

Stamens united by their anthers, and
  Not by their filaments. Flowers in heads, with a calyx-like involucre,   COMPOSITE F. 164
  Also generally by their filaments, more or less. Flowers not in heads.
    Corolla irregular, split down one side. Flowers perfect,   LOBELIA F. 167
    Corolla regular, succulent vines, with tendrils. Flowers monœcious,   †GOURD F. 154
Stamens separate from each other, and
  Inserted on the corolla. Leaves opposite or whorled.
    Leaves opposite, without stipules. Head of flowers with an involucre,   TEASEL F. 164
    Leaves opposite, without stipules. Head, if any, without an involucre.
      Stamens two or three fewer than the 5 lobes of the corolla,   VALERIAN F. 164
      Stamens as many as the lobes of the corolla, or one fewer,   HONEYSUCKLE F. 161
        Here one might expect to find the   †MIRABILIS F. 191
    Leaves whorled, without stipules, ⎫
    Leaves opposite, with stipules,   ⎬   MADDER F. 163
  Inserted with, but not on, the regular corolla.
    Stamens as many as the lobes of the corolla. Herbs,   CAMPANULA F. 167
    Stamens twice as many as the lobes of the corolla. Woody plants, ·   HUCKLEBERRY F. 168

B. COROLLA ON THE RECEPTACLE BELOW THE OVARY, i. e. Calyx free (except in Brookweed).

    1. *Stamens more in number than the lobes of the corolla.*

Leaves compound: pod one-celled. Flowers commonly irregular.
  Stamens 10 or rarely more when the flower is regular,   †PULSE F. 141
  Stamens 6 in two sets. Petals 4, united,   †FUMITORY F. 123
Leaves simple or palmately divided. Stamens many, monadelphous in a tube,   †MALLOW F. 131
Leaves simple, undivided. Stamens united only at the bottom, or separate.
  Stamens very many, adhering to the base of the corolla, ·   †CAMELLIA F. 134
  Stamens on the corolla, twice or four times as many as its lobes,   EBONY F. 172
  Stamens separate from the corolla, twice as many as its lobes,   HEATH F. 168

    2. *Stamens just as many as the lobes of the regular corolla*, 5, 4, *or rarely* 6 *or* 7.

Stamens one opposite each division of the corolla.
  Styles 5: calyx a chaff-like cup: petals 5, almost distinct,   LEADWORT F. 173
  Style 1. (Petals sometimes almost distinct,)   PRIMROSE F. 173
Stamens alternate with the divisions or lobes of the corolla, 5 or rarely 4,
  Inserted on the receptacle,   HEATH F. 168
  Inserted on the corolla, but connected more or less with the stigma. Juice
        milky. Ovaries and pods 2 to each flower.
    Anthers lightly adhering to the stigma: filaments monadelphous,   MILKWEED F. 188
    Anthers only surrounding the stigma: filaments distinct,   DOGBANE F. 187
  Inserted on the corolla, free from the stigma.

Style none: stigmas 4 to 6: corolla very short, deeply cleft,   HOLLY F. 171
Style one, rarely 2, sometimes 2-cleft or 3-cleft.
   Ovary deeply 4-lobed, in fruit making 4 akenes.
     Stamens 4. Leaves opposite, aromatic,   SAGE OR MINT F. 178
     Stamens 5. Leaves alternate, not aromatic,   BORRAGE F. 181
   Ovary and pod one-celled: the seeds on the walls.
     Leaves lobed or cut. Style 2-cleft above,   WATERLEAF F. 182
     Leaves entire and opposite, or alternate, with the 3 leaflets entire,   GENTIAN F. 187
   Ovary and fruit with 2 or more cells.
     Stamens 4, long. Flowers in a close spike,   PLANTAIN F. 172
     Stamens 5. Pod or berry many-seeded.
      Flower not quite regular. Style entire,   • FIGWORT F. 175
      Flower quite regular: stamens all alike,   NIGHTSHADE F. 185
     Stamens 5. Pod few-seeded.
      Twining herbs. Seeds large,   CONVOLVULUS F. 184
      Erect or spreading herbs. Style 3-cleft at the top,   POLEMONIUM F. 183

3. *Stamens 2 or 4, always fewer than the lobes of the corolla or calyx.*

Corolla more or less irregular, mostly 2-lipped.
   Ovary 4-lobed, making 4 akenes. Stems square: leaves opposite, aromatic, SAGE OR MINT F. 178
   Ovary and fruit 4-celled and 4-seeded. Stamens 4, }
   Ovary one-celled, making one akene. Stamens 4, }   VERVAIN F. 177
   Ovary and pod one-celled, many-seeded on the walls. No green leaves,   BROOM-RAPE F. 174
   Ovary and pod 2-celled with many large and winged seeds, }
   Ovary and fruit irregularly 4–5-celled, with many large seeds, }   BIGNONIA F. 174
   Ovary and pod 2-celled, with many or few small seeds,   FIGWORT F. 175
Corolla regular. Stamens only 2. Woody plants.
   Corolla 4-lobed or 4-parted,   OLIVE F. 189
   Corolla 5-lobed, salver-shaped,   JESSAMINE F. 188

III. Apetalous Division. Corolla none: sometimes the calyx also wanting. (Those which are merely apetalous forms of the preceding divisions are marked †.)

A. FLOWERS NOT IN CATKINS, OR CATKIN-LIKE HEADS.

1. *Seeds many in each cell of the ovary or fruit.*

Calyx with its tube coherent to the 6-celled ovary,   BIRTHWORT F. 190
Calyx free from the ovary.
   Pod 5-celled, 5-horned,   Ditchwort in †STONECROP F. 156
   Pod 3-celled, or one-celled with 3 or more styles,   Carpetweed, &c. in †PINK F. 129
   Pod or berry one-celled and simple,   †CROWFOOT F. 112

2. *Seeds only one or two in each cell of the ovary or fruit.*

Pistils more than one to the flower, and separate from each other.
   Calyx present and petal-like. Stamens on the receptacle,   †CROWFOOT F. 112
   Calyx present; the stamens inserted on it. Leaves with stipules,   †ROSE F. 146

Pistil only one, either simple or formed of two or more with their ovaries united.

Styles 10.   Fruit a 10-seeded berry,   POKEWEED F. 191

Styles or stigmas 2 or 3.

Herbs with sheaths for stipules, and entire leaves,   BUCKWHEAT F. 192

Herbs with separate stipules, and compound or cleft leaves,   HEMP F. 196

Herbs without stipules, and

Without scaly bracts. Flowers small and greenish,   GOOSEFOOT F. 191

With scaly bracts around and among the flowers,   AMARANTH F. 192

Shrubs or trees, with opposite leaves. Fruit a pair of keys,   †MAPLE F. 140

Shrubs or trees, with alternate leaves and deciduous stipules.

Stamens on the throat of the calyx, alternate with its lobes,   BUCKTHORN F. 138

Stamens on the bottom of the calyx, ,   ELM F. 195

Style one: stigma 2-lobed. Fruit a key. Leaves pinnate,   Ash in †OLIVE F. 189

Style or sessile stigma one and simple.

Calyx tubular or cup-shaped, colored like a corolla.

Stamens 8, on the tube. Shrubs: leaves simple,   MEZEREUM F. 195

Stamens 4, on the throat. Herbs: leaves compound,   Burnet in †ROSE F. 146

Stamens 5 or less on the receptacle. Calyx imitating a monopetalous

funnel-shaped corolla: a cup outside imitating a calyx.

Herbs with opposite leaves,   MIRABILIS F. 191

Calyx of 6 petal-like sepals colored like petals: stamens 9 or 12: anthers opening

by uplifted valves. Aromatic trees and shrubs,   LAUREL F. 194

Calyx in the sterile flowers of 3 to 5 greenish sepals: stamens the same number.

Flowers monœcious or diœcious,   NETTLE F. 195

B. FLOWERS ONE OR BOTH SORTS IN CATKINS OR CATKIN-LIKE HEADS.

Twining herbs, diœcious : fertile flowers only in a short catkin,   Hop in the HEMP F. 196

Trees or shrubs.

Sterile flowers only in catkins. Flowers monœcious.

Leaves pinnate. Ovary and fruit (a kind of stone-fruit, without an involucre), WALNUT F. 197

Leaves simple. Nuts one or more in a cup or involucre,   OAK F. 197

Both kinds of flowers in catkins or close heads.

Leaves palmately veined or lobed.

Calyx 4-cleft, in the fertile flowers becoming berry-like. Mulberry, &c. in NETTLE F. 195

Calyx none: flowers in round heads,   PLANE-TREE F. 196

Leaves pinnately veined.

Flowers diœcious, one to each scale. Pod many-seeded,   WILLOW F. 199

Flowers monœcious, the fertile ones 2 or more under each scale,   BIRCH F. 199

Flowers only one under each fertile scale. Fruit one-seeded,   SWEET-GALE F. 198

SUBCLASS II. — GYMNOSPERMS.

Proper pistil none ; the ovules and seeds naked, on the bottom or inner face of an

open scale, as in Pines, or without any scale at all, as in Yew,   PINE FAMILY, 201

I. Polypetalous Division.

1. CROWFOOT FAMILY. Order RANUNCULACEÆ.

Herbs, or sometimes slightly woody plants, with a colorless juice, sharp or acrid to the taste. Parts of the flower all separate and distinct, and inserted on the receptacle. Petals often wanting or of singular shapes. Stamens many, or at least more than 12. Pistils many, or more than one (except in Larkspur, Baneberry, and Bugbane), and entirely separate, except in Fennel-flower, in fruit becoming akenes or pods, or sometimes berries. The leaves are generally compound, or much cut or parted, and without stipules.

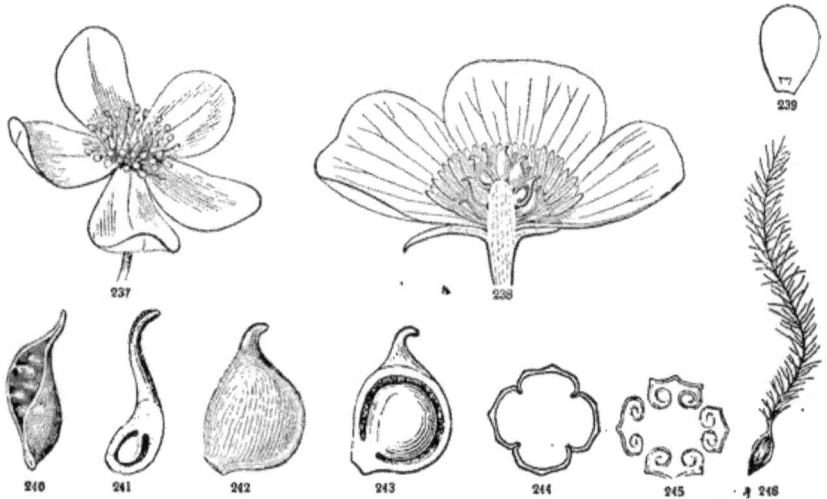

237. Flower of Pennsylvanian Anemony. 238. Half a flower of a Crowfoot, magnified. 239. A petal, showing its little scale. 240. Pod of Marsh-Marigold, opening. 241. A pistil of Anemony, magnified, the ovary cut through to show the ovule in it. 242 Akene of Crow-foot, enlarged. 243. Same, cut through to show the seed in it. 244 Enlarged cross-section of the segmts of Virgin's-Bower No. 1, in the bud. 245. Same of Virgin's-Bower No. 3. 246. Akene and feathery tail or style of Virgin's-Bower No. 1.

The genera are numerous. The following table or key leads to the name of each.

Climbing plants, with opposite, generally compound leaves, no real petals, the edges of
               the sepals turned inwards in the bud, (*Clématis*) VIRGIN'S-BOWER.
Not climbing: leaves all alternate except in Anemony: sepals overlapping in the bud.
    Pistils many or several, one-seeded, becoming akenes in fruit.
      Petals none: but the sepals colored like petals.
        Three leaves under the flower exactly imitating a calyx,   (*Hepática*) HEPATICA.

No such calyx-like leaves (or involucre) close to the flower.

Flowers single, on long, naked stalks,   *(Anemòne)* ANEMONY.

Flowers several in a simple umbel, handsome, }
Flowers many in a panicle, small,   }   *(Thalictrum)* MEADOW-RUE.

Petals present as well as sepals, the latter falling off early,   *(Ranunculus)* CROWFOOT.

Pistils more than one-seeded, becoming pods (except in Baneberry).

Sepals petal-like, not falling when the flower first opens, and

Golden-yellow: petals none. Leaves rounded, not cut,   *(Caltha)* MARSH-MARIGOLD.

Yellow or yellowish: petals stamen-like. Leaves deeply cut, *(Tróllius)* GLOBE-FLOWER.

White: pistils several, on stalks of their own. Leaflets 3,   *(Cóptis)* GOLDTHREAD.

Blue, purple, red, &c., rarely white. Pistils not stalked.

Pistils 5, united below into a bladdery pod,   **(Nigélla)** *FENNEL-FLOWER.

Pistils 2 to 5, rarely one, separate.

Sepals 5, all alike: petals 5, in the form of large spurs, *(Aquilègia)* COLUMBINE.

Sepals 5, dissimilar. Flower irregular.

Upper sepal long-spurred: petals 4,   *(Delphinium)* LARKSPUR.

Upper sepal hood- or helmet-shaped; petals 2,   *(Aconitum)* ACONITE.

Sepals petal-like, white, falling when the flower opens: petals minute or none.

Flowers in a short raceme. Pistil one, making a berry,   *(Actæa)* BANEBERRY.

Flowers in a long raceme. Fruit a dry pod,   *(Cimicifuga)* BUGBANE.

Sepals leaf-like, not falling off: petals large and showy,   *(Pæònia)* *PEONY.

*** Those genera which have more than one common species are next given, with the distinguishing marks of the species.

Virgin's-Bower. *Clèmatis.*

Calyx of 4 petal-like sepals, their margins not overlapping, but turned or rolled inwards in the bud. (Fig. 244, a cross-section of the calyx in the bud, shows this slightly in species No. 1, and Fig. 245, much rolled inwards, in No. 3.) No real petals. Fruit of many akenes, their style remaining generally in the form of a long and feathery tail (Fig. 246). Flowering in summer.

1. WILD VIRGIN'S-BOWER. Flowers white, in panicles, small, somewhat diœcious; leaflets 3, toothed; akenes with long feathery tails (Fig. 246). Banks of streams.   C. *Virginiàna.*

2. SWEET V. Flowers panicled, white; leaflets 5 to 9, entire. Cultivated in gardens.   C. *Flámmula.*

3. VINE-BOWER. Flower single; sepals purple, large; fruit short-tailed, naked. Cult.   C. *Viticélla.*

Hepatica (or Liverleaf). *Hepática.*

Calyx of 6 to 12 petal-like sepals, which are naturally taken for a corolla, because just underneath is a whorl of 3 little leaves exactly resembling a calyx; but it is a little way below the flower. Real petals none. Pistils several, making naked-pointed akenes. — Low herbs, in woods, sending up from the ground, in early spring, rounded 3-lobed leaves, which last over the next winter, and scapes with single (blue, purple, or nearly white) flowers.

1. ROUND-LOBED H. Lobes of the leaves 3, rounded and blunt. Common N. & E.   *H. triloba.*

2. SHARP-LOBED H. Lobes of the leaves 3 or 5, acute. Common W.   *H. acutiloba.*

Ane'mony. *Anemòne.*

Calyx of from 5 to 15 petal-like sepals; no leaves just underneath it, but the flowers on long and naked footstalks. No real petals. Akenes blunt or short-pointed, not ribbed nor grooved. Perennial herbs: their upper or stem-leaves opposite or in whorls. Flowers generally single, handsome. The following are the common wild species: they grow in woods and low meadows; the first three blossom in summer; the fourth in early spring.

1. VIRGINIAN ANEMONY. Principal stem-leaves 3 in a whorl, on long footstalks, 3-parted and cut-lobed, hairy; middle flower-stalk leafless, the others 2-leaved in the middle, new ones rising from their axils, and so producing the blossoms all summer; sepals greenish white, acute; pistils very many, in an oval woolly head. *A. Virginiàna.*

2. LONG-FRUITED A. Stem-leaves many in a whorl; flower-stalks 2 to 6, all leafless, very long; sepals blunt; head of fruit (an inch) long: otherwise like the last. N. & W. *A. cylindrica.*

3. PENNSYLVANIAN A. Hairy; stem-leaves sessile; main ones 3 in a whorl, but only a pair of smaller ones on each of the side flowering branches; sepals large, white or purplish; akenes flat, many in a round head. *A. Pennsylvánica.*

4. GROVE A. Smooth, low, one-flowered; stem-leaves 3 in a whorl, on long footstalks, divided into 3 or 5 leaflets; sepals white or purplish; akenes only 15 to 20, narrow. *A. nemoròsa.*

Meadow-Rue. *Thalictrum.*

Sepals 4 or more, petal-like or greenish. Real petals none. Pistils 4 to 15, becoming ribbed or grooved akenes. — Perennials, with compound leaves. No. 1 is almost an Anemony, except for its ribbed akenes, and has a few handsome and perfect flowers in an umbel. The other two have small and mostly diœcious flowers in a compound panicle, and decompound leaves; one of the lower leaves is shown in Fig. 133.

1. ANEMONY M. Low, delicate; stem-leaves all in a whorl at the top; sepals 7 to 10, white or pink-ish, like those of Grove Anemony, with which it generally grows. Fl. spring. *T. anemonoìdes.*

2. EARLY M. Plant 1° or 2° high; leaves all alternate, the rounded leaflets with 5 to 7 roundish lobes; flowers greenish, in early spring. Woods. *T. dioìcum.*

3. LATE M. Much like the last, but 3° to 6° high; leaflets 3-lobed; flowers white, in summer. Common in meadows and along streams. *T. Cornùti.*

Crowfoot. *Ranùnculus.*

Sepals 5, falling early. Petals 5 (sometimes accidentally more), flat. Akenes many in a head, flat.

    * Petals white, with a round spot at the base : herbage all under water.

1. WHITE WATER-CROWFOOT. Leaves made up of many delicate thread-like divisions. *R. aquátilis.*

    * * Petals yellow, and with a little scale on the inside at the bottom. (Fig. 239.)

    ← Herbage all or nearly all under water.

2. YELLOW WATER-C. Like the last, but larger in all its parts, and yellow-flowered, the upper leaves often out of water and much less cut. N. & W. *R. Púrshii.*

    ← ← Not growing under water.

    → Petals not longer, but often shorter, than the calyx : plants erect, in wet places.

3. SMALL-FLOWERED CROWFOOT. Very smooth, slender; first root-leaves crenate. *R. abortìvus.*

4. CURSED C. Very smooth, stouter; leaves all cleft or lobed; head of fruits oblong. *R. scelerátus.*

5. HOOK-BEAKED C. Hairy; leaves all 3-cleft, lobes broad; akenes with long and hooked beaks, collected into a round head. *R. recurvàtus.*

6. BRISTLY C. Stout, bristly-hairy; leaves divided into 3 or 5 stalked leaflets, which are cleft and cut again into narrow lobes; akenes straight-beaked, in an oblong head. *R. Pennsylvánicus.*

++ ++ Petals always much longer than the calyx. Dry ground, except No. 8.

7. EARLY C. Low, 4' to 9' high; root-leaves nearly pinnate; petals narrow. Fl. spring. *R. fascicularis.*

8. CREEPING C. Stems reclining, making long runners in summer; leaves variously divided; petals obovate. Wet places. *R. repens.*

9. BULBOUS C., or EARLY BUTTERCUP. A solid bulb at the base of the upright stem; leaves divided and cut; petals round, large, and bright yellow. Naturalized, E. in meadows. Fl. spring.
R. bulbosus.

10. TALL C., or LATER BUTTERCUP. Stem upright, 2° or 3° high, no bulb at the bottom; leaves divided and cut; petals obovate, not so large and bright-colored as the last. Fl. summer. *R. acris.*

Globe-flower. *Tróllius.*

Appears like a large Crowfoot or Buttercup, but the yellow leaves of the blossom are sepals ; within are the petals, small, and of peculiar shape, appearing like larger stamens. And the nine or more pistils make several-seeded pods.

1. EUROPEAN G. Sepals 10 to 15, golden-yellow, converging, and so making a rather globe-shaped flower; petals longer than the stamens. Cult. in gardens; fl. spring. *T. Europæus.*

2. AMERICAN G. Sepals 5 or 6, spreading, pale greenish-yellow; petals shorter than the stamens, and liable to be overlooked. Swamps, N. *T. Americánus.*

Columbine. *Aquilègia.*

Sepals 5, petal-like, all similar. Petals 5, in the form of large hollow spurs. Pistils 5, making many-seeded pods. — Leaves twice or thrice compound; leaflets in threes. (Fig. 247.)

1. WILD C. Flowers scarlet, yellow inside, nodding ; spurs hooked. Rocks. *A. Canadénsis.*

2. GARDEN C. Flowers blue, purple, or white; spurs straight. In all gardens. *A. vulgàris.*

Larkspur. *Delphinium.*

Sepals 5, petal-like, dissimilar, the upper one prolonged behind into a hollow spur. Petals 4, small; the upper pair with hardly any claws, but with long spurs which run back into the spur of the calyx: the lower pair with short claws and no spur ; in some species all the petals grow together into one body. Pistils and pods 1 to 5, many-seeded. — Flowers showy, in racemes or panicles, mostly white, blue, or purple. (Fig. 251, 252.)

* Garden annuals: leaves finely cut: petals united into one body (Fig. 253): pistil only one.

1. COMMON or FIELD LARKSPUR. Flowers scattered on spreading branches; pods smooth. *D. Consólida.*

2. ROCKET or AJAX L. Flowers crowded in a long and close raceme; pods hairy. *D. Ajàcis.*

* * Garden perennials : pistils 2 to 5 : the four petals separate. Many varieties are cultivated, mostly of the two following species.

3. GREAT-FLOWERED L. Leaves cut into linear distant lobes; pods downy. *D. grandiflòrum.*

4. BEE L. Leaves cleft into 3 to 7 wedge-shaped and cut-toothed lobes; petals bearded. *D. elàtum.*

* * * Wild species at the West and South: perennials, with 4 separate petals and 3 to 5 pods.

5. TALL WILD L. Stem 2° to 5° high; leaves parted into 3 or 5 narrow wedge-shaped pointed divisions; flowers many in a long raceme, blue-purple, in summer. *D. exaltàtum.*

6. DWARF L. Stem 1° high or less ; the 5 divisions of the leaves cleft into linear lobes; flowers few, loose, and large, purple-blue, in spring; pods spreading. *D. tricórne.*

7. AZURE L. Leaves parted and cut into narrow linear lobes; flowers many in a close raceme, sky-blue or white; pods erect.   *D. azùreum.*

Aconite. (Monkshood, Wolfsbane.) *Aconìtum.*

Sepals 5, petal-like, dissimilar, the upper one largest and forming a hood or *helmet.* Petals only 2, and those are small and curiously shaped bodies, with a curved or hammer-shaped little blade on a long claw, standing under the hood. Pods as in Larkspur. — Flowers in racemes or panicles, showy, blue, or purple, varying to white. Herbage and roots *poisonous.* (Fig. 254, 255.)

1. GARDEN ACONITE. Stem erect and rather stout, very leafy; divisions of the leaves parted into linear lobes; flowers crowded.   *A. Napellus.*
2. WILD A. Stem weak and bending, as if to climb; lobes of the leaves lance-ovate; flowers scattered, in summer. W.   *A. uncinàtum.*

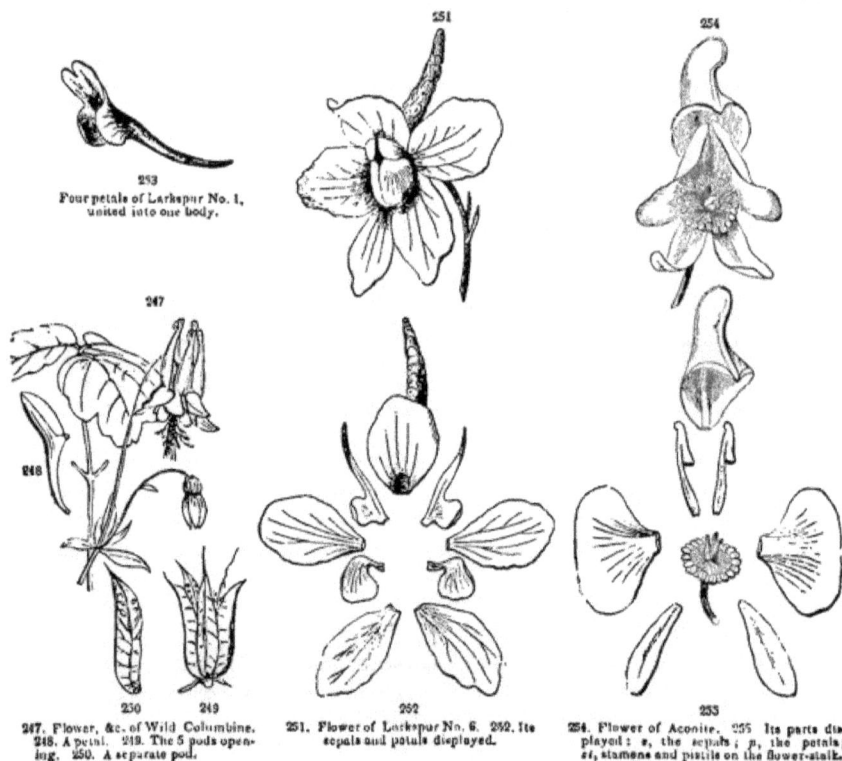

253. Four petals of Larkspur No. 1, united into one body.

247. Flower, &c. of Wild Columbine. 248. A petal. 249. The 5 pods opening. 250. A separate pod.

251. Flower of Larkspur No. 6. 252. Its sepals and petals displayed.

254. Flower of Aconite. 255. Its parts displayed: *s,* the sepals; *p,* the petals; *st,* stamens and pistils on the flower-stalk.

2. **MAGNOLIA FAMILY.** Order MAGNOLIACEÆ.

Trees or shrubs, with aromatic or strong-scented and bitter bark, and alternate simple leaves, which are never toothed; large, thin stipules form the covering of the buds, but fall off early. Flowers large, single at the ends of the branches; their leaves in threes, viz. 3 sepals colored like the petals, and 6 petals in two ranks or 9 in three ranks, their margins overlapping in the bud. Stamens very many, on the receptacle, with long anthers occupying, as it were, the side of the filament. Pistils many, packed and partly grown together one above the other, so as to make a sort of cone in fruit. — We have only two genera.

257

1. Stipules flat, not adhering to the leafstalk. Petals 6, greenish-or-ange. Filaments slender. Pistils overlying each other and grown to-gether to make a spindle-shaped cone, dry when ripe, and sepa-rating into a sort of key-fruit. Leaves somewhat 3-lobed, and as if cut off at the end. One species only is known, the

(*Liriodéndron Tulipifera*) TULIP-TREE.

2. Stipules making a round and pointed bud, adhering to the lower part of the leaf-stalk. Petals 6 to 9. Fil-aments below the anther very short. Cone of fruit rose-red and fleshy when ripe, the pistils opening on the back, the scarlet fleshy-coated seeds hanging by delicate and very elastic threads,   MAGNOLIA.

256. Small Laurel-Magnolia. 257. A stamen magnified. 258. Its cone of fruit, the seeds hanging as they drop.

Magnolia. *Magnólia.*

Our wild species divide into Laurel-Magnolias, Cucumber-trees, and Umbrella-trees.

§ 1. LAUREL-MAGNOLIAS. Leaves thick, evergreen at the South; leaf-buds silky; flowers rather globe-shaped, appearing through the summer, white, very fragrant

1. GREAT LAUREL-MAGNOLIA. Tree with leaves deep-green and shining above, rusty beneath when young; flower very large. S. It has stood the winter as far north as Philadelphia. *M. grandiflòra.*

2. SMALL LAUREL-M. (or WHITE BAY). Shrub or small tree; leaves oblong, whitish beneath; flower about 2' broad. Swamps. E. & S: *M. glauca.*

§ 2. CUCUMBER-TREES. Leaves thin, scattered along the branches, a little downy beneath, buds silky; flowers not sweet-scented, nor showy, nor very large, appearing in spring.

3. COMMON CUCUMBER-M. A tall tree; leaves oval or oblong, pointed; flowers greenish; young fruit resembling a very small cucumber. Common W.   *M. acumináta.*

4. YELLOW CUCUMBER-M. A low tree; leaves ovate or a little heart-shaped; flowers cream-yellow. S.; sometimes cultivated at the North.   *M. cordàta.*

§ 3. UMBRELLA-TREES. Leaves thin, large, those on the flowering shoots forming an umbrella-like circle underneath the blossom; leaf-buds smooth; flower large and white, not sweet-scented, appearing in early spring; petals about 4' long, tapering below.

5. EAR-LEAVED UMBRELLA-M. Leaves nearly 1° long, auricled at the base (Fig. 102). S.   *M. Fràseri.*

6. COMMON UMBRELLA-M. Leaves 1° to 2° long, tapering into a short footstalk.   *M. Umbrèlla.*

7. There is, besides, the GREAT-LEAVED M., with much the largest flowers and leaves of all, the latter 2° or 3° long, scattered, heart-shaped at the base, and white-downy beneath; flower 8' or 10' broad. S. and cult. rarely. It does not belong exactly to either the above divisions.   *M. macrophýlla.*

8. The PURPLE MAGNOLIA, from Japan, is a shrub in some gardens and grounds, flowering before the leaves are out.   *M. purpùrea.*

3. CUSTARD-APPLE FAMILY. Order ANONACEÆ.

Trees or shrubs, resembling the Magnolia family, but the three petals of each set not overlapping each other in the bud; the bark and foliage not aromatic, but unpleasant-tasted; the seeds large and bony, their albumen variegated like a nutmeg, or cut into slits. Leaves entire, destitute of stipules. Only one genus in this country, and one species common; the

1. COMMON PAPAW. A small tree, with dingy-purple flowers appearing in early spring rather before the leaves; the 3 outer petals much larger than the 3 inner ones; fruits eatable when ripe, in autumn, 2' or 3' long. Common West and South along rivers, in rich soil.
     Asimina trìloba.

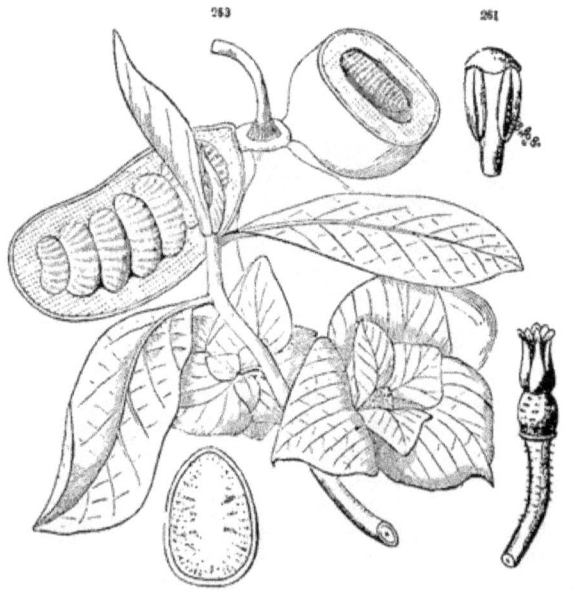

260 Branch of Papaw in flower. 261, A stamen. 262, Flower with all but the pistils taken off the receptacle. 263, Fruits; two of them cut through. 264, A seed cut through to show the variegated albumen.

4. MOONSEED FAMILY. Order MENISPERMACEÆ.

Woody climbers, with alternate leaves and small diœcious flowers (as shown in Fig. 167, 168) ; the sepals and petals each 4 or 6 and both of the same color, and a few one-seeded pistils, becoming small drupes in fruit, with a moon-shaped or kidney-shaped stone. We have two genera of one species each, the first common at the North, the second at the South.

1. Stamens 12 to 20 : pistils 2 to 4. Flowers white : leaves rounded and angled shield-shaped. Fruit blue-black, (*Menispérmum*) MOONSEED.

2. Stamens 6, one before each petal. Flowers greenish: leaves heart-shaped. (*Cócculus*) COCCULUS.

5. BARBERRY FAMILY. Order BERBERIDACEÆ.

Readily distinguished (with a single exception) by having the sepals and petals in fours, sixes, or eights (not in fives), and with just the same number of stamens as petals, *one before each petal* (on the receptacle), the anthers opening by an uplifted valve or door on each side. Pistil only one. Harmless, except the May-Apple (also called MANDRAKE), which has rather poisonous roots, although the fruit is innocent and eatable. Having only one species of each genus, we may ascertain them by the following key : —

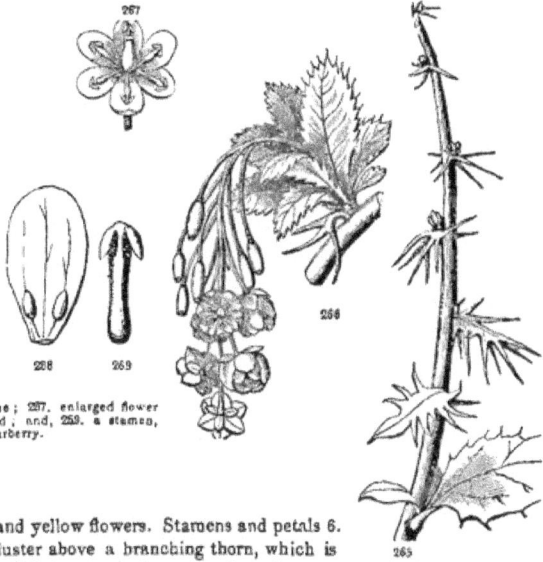

266. Shoot ; 268. cluster of leaves and raceme ; 267. enlarged flower spread open ; 266. a petal more magnified ; and, 269. a stamen, with the anther opening, of the common Barberry.

Shrubs with yellow bark and wood, and yellow flowers. Stamens and petals 6.
  Leaves appearing simple, in a cluster above a branching thorn, which is
          an altered leaf of the year before. Berries red, (*Bérberis*) BARBERRY.
  Leaves scattered, pinnate, evergreen: no thorns. Berries blue, (*Mahónia*) * MAHONIA.
Herbs, with perennial roots, all with compound or deeply lobed leaves.
  Flowers yellowish-green, small. Stamens and petals 6. Leaves decompound, from
          the root and also at the top of the stem, (*Caulophyllum*) COHOSH.

Flowers white, rather large: petals larger than the fugacious calyx.

  Stamens and narrow petals 8.   The one-flowered scape and the 2-parted leaves rising separately from the ground. Fruit a many-seeded pod opening by a lid,   (*Jeffersònia*) TWINLEAF.

  Stamens 12 to 18: petals rounded, 6 to 9.   Flowering stems 2-leaved at the top: leaves shield-shaped and several-cleft, large, with a nodding flower in the fork,   (*Podophýllum*) MAY-APPLE.

6. WATER-LILY FAMILY.   Order NYMPHÆACEÆ.

Water-plants with flowers and leaves on long footstalks, rising out of the water or rest-ing on its surface; the leaves either shield-shaped or deeply heart-shaped. Petals and sta-mens generally very many.— To the proper Water-Lily fam-ily may here well enough be added the Water-shield and the Nelumbo, each of a sin-gle species. This gives us four genera, which are distin-guished as follows:—

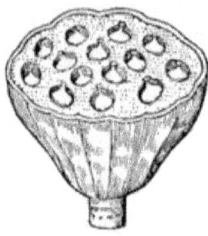

Fruit of Nelumbo.

270. Flower, bud, and leaf of White Water-Lily. 271 Flower with the parts cut away, all but two petal-like stamens, one ordinary stamen, and the compound pistil. 272. Slice across the 11-celled pistil.

1. Leaves and flowers from very thick and long creeping rootstocks. Sepals and the many petals and stamens gradually blending into each other, and growing to the surface of the many-celled and many-seeded compound pistil. Flower white, sweet-scented,   (*Nymphæa*) WHITE WATER-LILY.

2. Leaves and flowers from rootstocks like the last. Sepals 5 or 6, rounded, partly petal-like and yellow. Within these a mass of small, square-topped bodies looking like and not much larger than the stamens, but really answering to petals; and above them the real stamens in great numbers, all under the many-celled and flat-topped pistil, (*Nuphar*) YELLOW POND-LILY.

3. Leaves and small dull-purple flowers from a slender stem rising in the water; the oval leaves attached by the middle of the under side (centrally peltate). Sepals and petals narrow, each 4, and 12 to 18 stamens, all under the 4 to 16 separate and few-seeded pistils, (*Brasènia*) WATERSHIELD.

4. Sepals and petals (alike in many ranks) and stamens many, all falling off early, all under the pistils, which are 12 or more in number and separately embedded in the flat upper face of an enlarged top-shaped receptacle. In fruit they are round and eatable nuts (Fig. 273). Leaves very large (1° or 2° broad), round, attached by the middle underneath, cupped, rising out of the water, as do the great greenish-yellow flowers also, on long stalks. Common W. & S. (*Nelùmbium*) NELUMBO.

7. SIDESADDLE-FLOWER FAMILY. Order SARRACENIACEÆ.

Bog-plants with hollow, pitcher-shaped or trumpet-shaped leaves, all from the root, making the curious genus

Sidesaddle-Flower. *Sarracènia.*

Sepals 5, colored, persistent; and below the calyx are 3 small bractlets. Petals 5, fiddle-shaped, curved inwards. Stamens very many, on the receptacle. Style with a broad and 5-angled umbrella-shaped top, covering the 5-celled ovary and the stamens. Pod many-seeded. Flower single, large, nodding on the summit of a long scape.

1. PURPLE SIDESADDLE-FLOWER, or PITCHER-PLANT. Petals deep purple, arched over the pistil; leaves pitcher-shaped, yellowish-green, veined with purple, and with a broad wing down the inner side. Common N. & S. *S. purpùrea.*

2. RED S. Petals red; leaves long, trumpet-shaped, with a narrow side wing. S. *S. rùbra.*

3. SPOTTED S. Petals yellow; leaves trumpet-shaped, 12' to 18' long, with a hooded top spotted with white on the back, and a narrow side wing. S. *S. variolàris.*

4. YELLOW S., or TRUMPETS. Petals yellow, drooping when old; leaves 1° to 3° long, trumpet-shaped, with an upright rounded top turned back at the sides, side wing hardly any. Very common S. *S. flava.*

273
Leaves of Purple S.; one of them cut off.

8. POPPY FAMILY. Order PAPAVERACEÆ.

Herbs with a milk-white, yellow, or reddish juice (colorless in Eschscholtzia), which is bitter or acrid and poisonous, alternate leaves, and flowers remarkable for having only 2 (rarely 3) sepals, which fall when the blossom opens, but 4 (or in one case 8 or 12) petals, which fall early. Stamens many, on the receptacle. Pistil one, compound, but almost always one-celled, many-seeded, the seeds borne on the walls or on projections from them (parietal). Eschscholtzia is remarkable for its calyx shaped like a pointed cap or a candle-extinguisher. — In most cases we have only one species of each genus.

275. A flower-bud casting its calyx, and, 276, a flower of Poppy.

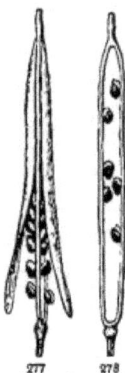

277. Pod of Celandine opening. 278. Frame of the same, turned flatwise, and seeds still on it.

279. Flower-bud, &c. of Eschscholtzia. 280. The cap-shaped calyx fallen off. 281. The pod.

Petals 4, crumpled or plaited in the bud, which nods before opening (except in the Prickly-Poppy).
    Ovary and pod incompletely several-celled, by plates or placentas projecting from
         · the walls and covered with numberless seeds. Stigmas making a flat
         sessile cap. Pod hard, opening by pores under the edge of the cap of
         stigmas,   (*Papaver*)   ✱ POPPY.
    Ovary and pod strictly one-celled, opening by valves, and leaving the placentas as a
         slender frame between them. Flowers yellow, rarely white.
         Pod and leaves prickly. Style none: stigmas 4 or 6,   (*Argemòne*) PRICKLY-POPPY.
         Pod bristly. Style present: stigmas 3 or 4,   (*Stylóphorum*) CELANDINE-POPPY.
         Pod smooth, slender (Fig. 277): stigmas 2,   (*Chelidònium*) CELANDINE.
    Ovary and long narrow pod 2-celled by a thick partition in which the seeds are em-
         bedded; stigma 2-horned,   (*Glaucium*) ✱ HORN-POPPY.
Petals not crumpled in the bud, which does not nod.
    Petals 8 to 12, narrow, white. Pod oblong. Juice orange-red,   (*Sanguinària*) BLOODROOT.
    Petals 4, broad, yellow. Sepals united into a pointed cap which falls off as a lid (Fig.
         280, 281). Receptacle or end of the flower-stalk expanded and top-
         shaped. Stigmas 3 to 7, slender, unequal. Pod many-ribbed. Juice
         watery, colorless, but strong-scented,   ✱ ESCHSCHOLTZIA.

9. FUMITORY FAMILY. Order FUMARIACEÆ.

Tender herbs with a colorless juice, compound alternate leaves, and irregular flowers with only two small scale-like sepals, a flattened and closed corolla of 4 petals more or less grown together, the two outside ones larger with small spreading tips, the two inner small and with spoon-shaped tips sticking together face to face over the anthers and stigma: stamens on the receptacle, 6 in two sets or bundles, one before each of the larger petals, or all joined in one tube below. The middle anther of each set is two-celled; the side ones only one-celled. Pistil one, in the manner of the Poppy family. Pod one-celled. Bitterish, harmless plants, with singularly shaped flowers, some of them handsome. We have four genera, two of them of only one species each.

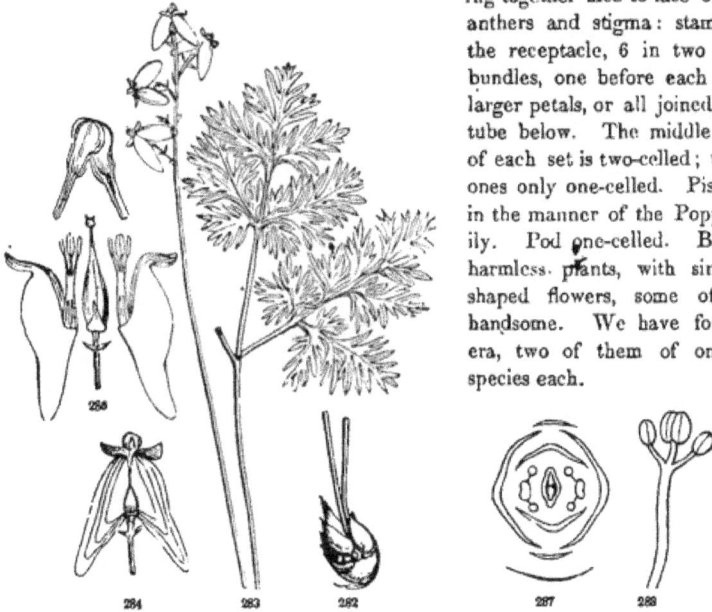

282. Bulb, and, 283, leaf and flowers of Dicentra No 1. 284. Flower, natural size. 285, 286. Same, taken to pieces. 287. Diagram of the flower of a Corydal. 288. One of the sets of stamens united.

Flower heart-shaped, or with a spur on each side at the base.
  Petals all permanently united into a slightly heart-shaped (pale flesh-colored) corolla, which dries without falling and encloses the four-seeded pod. A delicate vine climbing by the tendril-like divisions of its thrice-pinnate leaves, (*Adlùmia*) SMOKE-VINE.
  Petals less united, readily separated. Pod several-seeded, (*Dicéntra*) DICENTRA
Flower with a projection or spur at the base on one side only.
  Ovary slender, forming a several-seeded pod, (*Corýdalis*) CORYDAL.
  Ovary and fruit, round, small, one-seeded, not opening, (*Fumària*) FUMITORY.

9

Dicentra. *Dicéntra* (wrongly called *Dielytra*).

The species are perennials with singular and handsome flowers in racemes, blossoming in spring. Wild species, in rich woods; the decompound and finely cut leaves and naked flower-stalk rising separately from the ground, in early spring. Delicate low plants, chiefly found N. & W.

1. Dutchman's Breeches D. (Fig. 282–286.) Herbage from a sort of bulb of coarse grains; corolla white, tipped with cream-color, with 2 very large spurs. *D. Cucullària.*

2. Squirrel-Corn D. Underground shoots bearing little yellow tuber-like bodies, resembling grains of Indian Corn; corolla white and flesh-color, fragrant like Hyacinths. *D. Canadénsis.*

  * * Garden species, leafy-stemmed, 2° or 3° high, with Peony-like leaves.

3. Snowy D. Racemes drooping, one-sided; flowers pink-purple, 1' long. Cultivated. *D. spectábilis.*

Corydal. *Corýdalis.*

Our two species are leafy-stemmed biennials, glaucous, with twice-pinnate leaves, and linear or slender pods. They grow in rocky places and flower in spring and summer.

1. Golden C. Low and spreading; flowers yellow in simple racemes; pods hanging. *C. aúrea.*

2. Pale C. Upright; flowers purplish and yellowish; racemes panicled; pods erect. *C. glauca.*

10. CRUCIFEROUS OR CRESS FAMILY. Order CRUCIFERÆ.

Herbs, with alternate leaves, a sharp-tasted watery juice (never poisonous, but often very acrid or biting); perfectly distinguished by their *cruciferous* flowers, *tetradýnamous* stamens, and by having the sort of pod called a *siliq̇ue* or *silicle* (240, 241). The flower is called cruciferous because the 4 petals, with claws enclosed in the 4-sepalled calyx, have their blade spreading so as to form the four arms of a cross. As to the stamens, they are 6 in number (on the receptacle), two of them always shorter than the other four. The pistil makes a pod, like that of the Celandine, &c. in the Poppy family (Fig. 277), except that a partition stretches across between the two thread-shaped placentas, and divides the cavity into two cells. When the pod opens, the two valves fall away, leaving the seeds attached to the edges of this frame. The whole kernel of the seed is an embryo. It is always bent or folded up, in various ways. The flowers of the whole family are so much alike, that the genera have to be distinguished by their pods and seeds. This makes the family too difficult for the beginner. But so many plants of the family are common in cultivation, that we add a tabular key, leading to the names of the principal kinds.

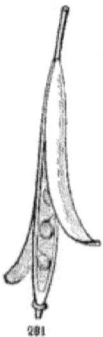

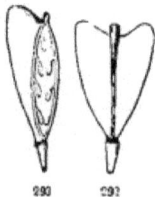

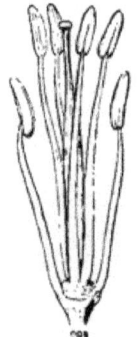

289. Flower of Mustard. 290. Stamens and pistil, more magnified. 291. Pod (siliq̇ue) of Toothwort, opening. 292. Pod (silicle or pouch) of Shepherd's-Purse. 293. Same, with one valve fallen off.

1. *Pod (silique) generally several times longer than wide.*

Pod not splitting open when ripe, but becoming hard, beak-pointed. Seeds round.

  Flowers pink or purple. Pod thick, fleshy when young,   (*Ráphanus*) *RADISH.

  Flowers yellow, turning whitish or purplish. Pod long, necklace-shaped,

                        (*Raphanus, § Raphanistrum*) JOINTED-CHARLOCK.

Pod splitting, i. e. opening when ripe by two valves, which fall off and leave the partition.

  Pod ending in a beak. Seeds round. Flowers yellow.

    Calyx erect in blossom.   Roots, stems, or leaves, &c. be- ⎫ (*Brássica*) ⎰*TURNIP and
                        coming fleshy in cultivation, ⎭   ⎱*CABBAGE.

    Calyx open or spreading in blossom,   (*Sinapis*) MUSTARD.

  Pod not beaked, i. e. not ending in a strong-pointed tip. Seeds flat or oblong.

    Calyx unequal, two of the sepals projecting or pouch-shaped at the base.

      Flowers yellow or orange. Pod and seeds flat,   (*Cheiránthus*) *WALLFLOWER.

      Flowers rose, purple, or white. Pods not flat.

        Stigmas thickened on the back. Seeds flat,   (*Matthiola*) *STOCK.

        Stigmas close-pressed together. Seeds oblong,   (*Hésperis*) *ROCKET.

    Calyx equal, i. e. the sepals all alike or nearly so.

      Pods flat. Flowers white or purple.

        Valves of the pod with a mid-nerve or vein,   (*Árabis*) ROCK-CRESS.

        Valves of the pod without a nerve.

          Stem-leaves alternate or scattered,   (*Cardámine*) BITTER-CRESS.

          Stem-leaves 2 or 3, whorled or clustered. Root fleshy, (*Dentária*) TOOTHWORT.

      Pods obtusely 4-sided, linear. Flowers yellow,   (*Barbarèa*) WINTER-CRESS.

      Pods awl shaped. Flowers pale yellow,   (*Sisýmbrium*) HEDGE-MUSTARD.

      Pods turgid, short-linear or oblong,   (*Nastúrtium*) WATER-CRESS.

2. *Pod (silicle or pouch) short, the length not more than two or three times the breadth.*

Pod opening when ripe by 2 valves which fall off and leave the partition.

  Pod globose or ovoid, many-seeded,   (*Armorácia*) HORSERADISH.

  Pod pear-shaped, rather flattish, many-seeded. Flowers yellow,   (*Camelina*) FALSE-FLAX.

  Pod flat, with a broad partition. Seeds many,   (*Draba*) WHITLOW-GRASS.

  Pod flat, with a broad partition. Seeds 2 to 4.

    Flowers purple, large. Pod large, stalked above the calyx,   (*Lunária*) *HONESTY.

    Flowers white, small. Pod small, 2-seeded,   (*Kóniga*) *SWEET-ALYSSUM.

  Pod flattened contrary to the narrow partition. Flowers white or purple.

    Seeds many; pod triangular-obcordate with a shallow notch, (*Capsélla*) SHEPHERD'S-PURSE.

    Seeds only one in each cell.

      Petals all alike. Flowers very small,   (*Lepídium*) PEPPERGRASS.

      Petals unlike; the two on the outer side of the flower larger,   (*Ibéris*) *CANDYTUFT.

Pod not opening, 1-celled, 1-seeded, wing-like. Flowers yellow,   (*Isátis*) *WOAD.

Pod not opening, but jointed across the middle, fleshy. Flowers purplish,   (*Cakile*) SEA-ROCKET.

11. MIGNONETTE FAMILY. Order RESEDACEÆ.

    A family consisting of a few European herbs, with small and irregular flowers, which deserves notice merely because it contains the

Mignonette. *Resèda.*

Sepals 4 to 7, green, not falling off, open in the bud. Petals 4 to 7, unequal, on broad claws, the small blade cleft as if cut into several narrow slips. Stamens 10 or more, borne on an enlargement of the receptacle, turned to one side of the blossom. Pod short and broad, one-celled, dividing at the top into 3 to 6 horns, opening between the horns long before the seeds are ripe. The seeds are kidney-shaped, numerous, and parietal, that is, borne along the walls of the pod. — Herbs, with alternate leaves and small dull-looking flowers crowded in a raceme or spike.

1. COMMON MIGNONETTE. Low and spreading; leaves some entire, others 3-cleft; sepals and petals 6 or 7. Cultivated for its very fragrant small flowers. *R. odoràta.*

2. DYER'S-WEED. Stem simple, upright, 2° high; leaves all entire, broadly lance-shaped; sepals and petals 4. A weed along road-sides in some places; used for dying yellow. *R. Lutèola.*

12. VIOLET FAMILY. Order VIOLACEÆ.

Herbs with 5 sepals, 5 petals, and 5 stamens borne on · the receptacle, the lower petal rather different from the rest and enlarged at the bottom into a projecting sac or spur. Stamens very short and broad, the anthers a little united by their edges around the pistil. Pistil one, with one style. Pod one-celled, with three rows of seeds on its walls. — Leaves with stipules. Roots and juice rather acrid. The common plants of the family belong to the genus,

1. Violet. *Viola.*

Flower nodding on the summit of the flower-stalk. Style club-shaped; stigma bent over to one side. — Flowering in spring, and some species continuing to blossom all summer.

＊ Stemless species, i. e. leaves and naked flower-stalks all from rootstocks on or under ground.

← Garden species from Europe spreading by runners or rootstocks above ground.

1. SWEET OR ENGLISH VIOLET. Leaves rounded heart-shaped; flowers blue-purple, also a white variety, very fragrant. Cultivated, generally double-flowered. *V. odoràta.*

← ← Wild species, with tufted and fleshy uneven rootstocks. Flowers short-spurred.

↔ Flowers purple or blue, nearly scentless.

2. COMMON BLUE V. Flowers pretty large; side-petals bearded; leaves on long upright stalks, heart-shaped or kidney-shaped, the sides at the bottom rolled in when young, slightly toothed, or in the lobed or *Hand-leaf* variety cleft or parted in various degrees. Low grounds. *V. cucullàta.*

3. HAIRY V. Leaves short-stalked and flat on the ground; flowers smaller; otherwise like the last. Dry soil, S. & W. *V. villòsa.*

4. ARROW-LEAVED V. Early leaves on short and margined footstalks, oblong-heart-shaped, halberd-shaped, arrow-shaped, lance-oblong or ovate. Varying greatly, hairy or smoothish; side petals or all of them bearded; flowers large for the size of the plant. Dry or moist ground. *V. sagittàta.*

5. BIRD'S-FOOT V. Leaves cut into fine linear lobes; petals lilac-purple, large, beardless. Moist sandy soil. *V. pedàta.*

↔ ↔ Flowers small, white, faintly sweet-scented, the lower petal streaked. Small, in damp soil.

6. BLAND V. Leaves rounded heart-shaped or kidney-shaped; petals without any beard. *V. blanda.*

7. PRIMROSE-LEAVED V. Leaves oblong or ovate; side-petals generally bearded. *V. primulæfòlia.*

8. LANCE-LEAVED VIOLET. Leaves lance-shaped, erect, smooth; petals not bearded. *V. lanceolàta.*

++ ++ ++ Flowers light yellow, small.

9. ROUND-LEAVED V. Leaves round ovate and heart-shaped, spreading flat on the ground; side-petals bearded and brown-streaked inside. Cold woods, N. *V. rotundifòlia.*

* * Leafy-stemmed species.

← Wild species, perennial, with heart-shaped leaves, blossoming nearly all summer.

10. LONG-SPURRED V. Spur ½' long, considerably longer than the pale bluish corolla. *V. rostràta.*

11. MUHLENBERG'S V. Low, spreading by runners; spur stout, not more than half the length of the pale violet corolla. Wet woods. *V. Muhlenbérgii.*

12. PALE V. Spur much shorter than the cream-colored corolla; lower petal streaked. *V. striàta.*

13. CANADA V. Tall; petals white above, violet-tinged beneath; spur very short. *V. Canadénsis.*

14. DOWNY V. Tall, leafless below, downy; corolla yellow, spur very short. *V. pubéscens.*

← ← Cultivated or run wild; root annual or biennial.

15. HEART'S-EASE OR PANSY V. Low; upper leaves oval, the lower heart-shaped; stipules large and leaf-like, pinnatifid; corolla yellow-whitish, violet-blue, and purple, varying or mixed, large and showy in the cultivated Pansy, becoming small when run wild. *V. tricolor.*

13. CISTUS FAMILY. Order CISTACEÆ.

This small family consists of low shrubby plants or perennial herbs, in Europe with a showy corolla which opens only once, in sunshine, the petals falling off at sunset. Here it contains only a few less handsome, or homely, weed-like plants. They may be known by the following marks. — Leaves some of them alternate. Calyx remaining after blossoming, of 5 sepals, three of them large and two smaller, often very small, the latter entirely outside in the bud and looking like bracts. Petals 5 or 3, all alike, overlapping each other in the bud, each with one edge covering the one before it, but covered by that behind it. Stamens from 3 to 20 or more, all separate, borne on the receptacle. Pistil one, making a one-celled pod, with the seeds borne on three lines down the walls, or on projections from them, that is, with 3 parietal placentas.

294. Flower, &c. of Frostweed. 295. Its calyx and pistil. 296. Its ovary cut across and magnified.

Petals 5, yellow, falling after the flower has opened for one day only.
Style none. Petals crumpled in the bud, (*Heliánthemum*) FROSTWEED.
Style slender. Little shrubs with minute leaves on sandy shores, (*Hudsònia*) HUDSONIA.
Petals 3, purplish, persistent. Flowers very small. (*Léchea*) PINWEED.

14. ST. JOHN'S-WORT FAMILY. Order HYPERICACEÆ.

Herbs or low shrubs, with the leaves all opposite and dotted, as if punctured, with transparent or dark-colored dots, one or both; the juice generally acrid. Flowers with 4 or 5 persistent sepals, as many petals, and more numerous, commonly a great number of stamens, and in 3 or 5 clusters, borne on the receptacle. Styles 2 to 5, commonly separate, or sometimes all united into one. Ovary only one, in fruit a pod, either one-celled with 2 to 5 (commonly 3) parietal placentas, or with as many cells and the placentas in the inner angle of each cell (Fig. 189, 190), when ripe splitting through the partitions (Fig. 212).

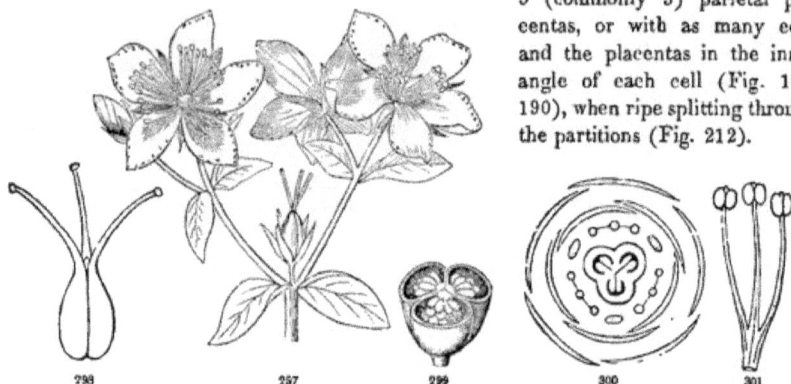

297. Flowers, &c. of St. John's-wort No. 4. 298. Pistil of 3 united. 299. Pod cut across. 300. Plan of the flower of Marsh St. John's-wort, in a cross section of the bud. 301. One of the clusters of three stamens.

Sepals 5, all nearly alike in size and shape.
　　Petals 5, flesh-colored, oblong, equal-sided, stamens about 9, in three sets, and a thick
　　　　　　gland between each set,　　　　　　　(*Elodèa*) MARSH ST. JOHN'S-WORT.
　　Petals 5, yellow, unequal-sided. Stamens generally many,　　(*Hypèricum*) ST. JOHN'S-WORT.
Sepals 4, in two pairs, one pair large, the other small; petals 4,　(*A'scyrum*) ST. PETER'S-WORT.

St. John's-wort. *Hypèricum.*

　　＊ Stamens very many, in 5 sets. Styles 5, rarely 6 or 7.

1. GREAT ST. JOHN'S-WORT. Perennial herb, with stems branched, 3° to 6° high; leaves closely sessile, oblong; petals 1' long, narrow. N. & W.　　　　　　　　*H. pyramidàtum.*

　　＊ ＊ Stamens very many. Styles 3 or splitting into 3. Perennials or shrubs.

2. SHRUBBY S. Shrub 1° to 4° high, very bushy; branchlets 2-edged; leaves lance-oblong; styles at first all united into one (Fig. 190), when old splitting into three. W. & S.　　*H. prolificum.*

3. NAKED-FLOWERED S. Shrubby at the base, 1° to 4° high; branches sharply 4-angled; leaves oblong; cyme stalked and naked. S. & W.　　　　　　　　　　　*H. nudiflòrum.*

4. COMMON S. Herb 1° or 2° high, bushy-branched; stem somewhat 2-edged; leaves narrow-oblong, with transparent dots; sepals lance-shaped; petals bright yellow. A weed in pastures, &c.

　　　　　　　　　　　　　　　　　　　　　　　　　　　　　　　　H. perforàtum.

5. CORYMBED S. Herb 1° to 2° high, with a terete stem, little branched; leaves oblong, dotted with black as well as with transparent dots, and so generally are the pale yellow petals; sepals oblong. Low grounds. *H. corymbosum.*

* * * Stamens few, 5 to 15. Styles 3, short. Pod one-celled. Slender annuals, growing in wet or sandy places, 4' to 15' high: flowers very small.

6. SMALL S. Stem weak, with spreading branches, leafy to the top; leaves ovate or oblong, partly clasping, 5-ribbed. *H. mutilum.*

7. CANADA S. Branches erect, leaves lance-shaped or linear; cymes leafless. *H. Canadense.*

8. PINE-WEED S. Bushy-branched, the branches wiry and very slender; the leaves very minute, awl-shaped, close-pressed to the branches; flowers minute, sessile along the branches. *H. Sarothra.*

15. PINK FAMILY. Order CARYOPHYLLACEÆ.

Herbs with opposite and entire leaves, which are not dotted, the stems swollen at the joints. Flowers regular, their parts in fives, sometimes in fours. Stamens never more than twice as many as the petals or sepals, and often fewer, on the receptacle or the calyx. Styles or stigmas generally separate, 2 to 5. Fruit a pod, which is generally one-celled, with the seeds from the bottom or on a central column. These are kidney-shaped, and have the embryo on the outside of the albumen, generally coiled around it. — Bland

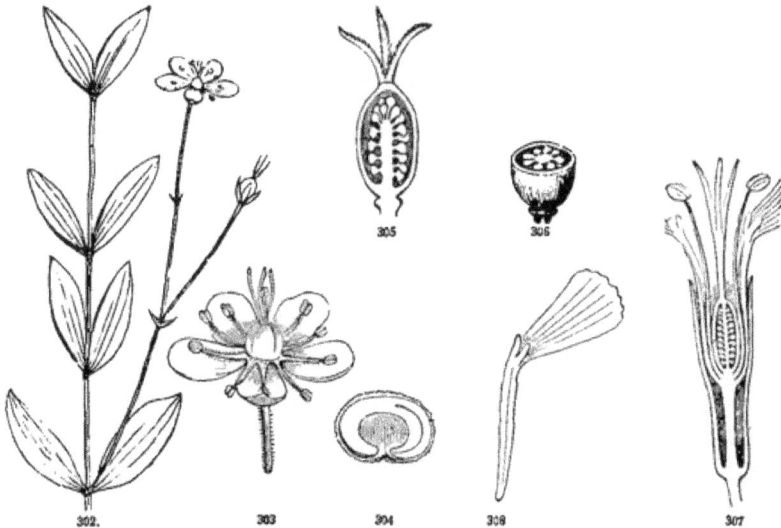

302. Piece of Side-flowering Sandwort. 303. Flower magnified. 304. A seed divided, showing the embryo coiled around the outside of the albumen. 305. Pistil of Sand-Spurrey cut through lengthwise and magnified. 303. Lower part of the ovary of the same, cut across. 307. Flower of a Catchfly cut through lengthwise. 308. A separate petal.

herbs: some are insignificant weeds; others have handsome flowers, and are cultivated for ornament. They form two main sub-families, one containing the Pinks, the other the Chickweeds. For lack of room, only the principal genera can be given here, without the species, which are numerous.

I. PINK SUBFAMILY. Sepals united into a tube or cup. Petals with long claws, which are enclosed in the tube of the calyx. The petals and the 10 stamens are generally raised more or less on a stalk within the calyx. Pod many-seeded, opening at the top. Flowers mostly rather large and showy.

Calyx furnished with two or more scaly bractlets at the base. Styles 2,	(*Diánthus*)	* PINK.
Calyx naked, i. e. without any bractlets at the base.		
Styles 2. Calyx cylindrical and even,	(*Saponária*)	SOAPWORT.
Styles 2. Calyx oblong and strongly 5-angled,	(*Vaccária*)	COWHERB.
Styles 3. Calyx 5-toothed,	(*Siléne*)	CATCHFLY.
Styles 5.		
Calyx with short teeth, which are not leaf-like,	(*Lychnis*)	* LYCHNIS.
Calyx with leafy lobes, which are longer than the petals,	(*Agrostémma*)	COCKLE.

II. CHICKWEED SUBFAMILY. Sepals separate or nearly so. Petals without claws, spreading, sometimes wanting. Small or low herbs; many are weeds. Flowers small, mostly white, except in Sand-Spurrey.

Pod 3-celled, many-seeded. Petals none. Prostrate annual weed,	(*Mollúgo*)	CARPETWEED.
Pod one-celled, with several or many seeds. Styles 3 to 5.		
Stipules or little scales between the leaves none.		
Petals 2-cleft or parted, or notched at the end.		
Styles and petals 5. Pod opening by 10 teeth.	(*Cerastium*)	MOUSE-EAR CHICKWEED.
Styles 3 or 4. Pod splitting into valves,	(*Stellária*)	CHICKWEED.
Petals entire, not notched nor cleft.		
Styles 3, fewer than the petals,	(*Arenária*)	SANDWORT.
Styles 5 or 4, as many as the petals,	(*Sagìna*)	PEARLWORT.
Stipules in the form of scales between the bases of the leaves.		
Styles 3. Leaves not whorled. Petals purple,	(*Spergulária*)	SAND-SPURREY.
Styles 5. Leaves in whorls, narrow. Petals white,	(*Spérgula*)	SPURREY.

16. PURSLANE FAMILY. Order PORTULACACEÆ.

More or less fleshy herbs, with entire leaves, and flowers which open only in sunshine. Sepals fewer than the petals (i. e. sepals 2, petals 5), with a stamen before each one, or else with many stamens. Pod one-celled, with the seeds, like those of the Pink family, on stalks rising from the base of the cell. Harmless and tasteless herbs; the Spring-Beauty has handsome flowers in the spring in woods. The common Purslane is a well-known garden weed and pot-herb, and the Great-flowered Purslane, with its cylindrical fleshy leaves and large red or scarlet flowers, is a common ornamental annual in cultivation.

Calyx 2-cleft, the tube united with the lower part of the ovary. Petals opening only once. Stamens 7 to 20. Pod many-seeded, opening round the middle, the top falling off as a lid. Annuals.   (*Portulàca*) PURSLANE.

Calyx 2-leaved, free from the ovary, which makes a few-seeded pod, splitting into 3 valves.
Stamens 5, one before each petal. Leaves 2 and opposite in our species, on a
stem which comes from a small tuber. Flowers rose-color, in a raceme, opening for several days. (*Claytónia*) SPRING-BEAUTY.

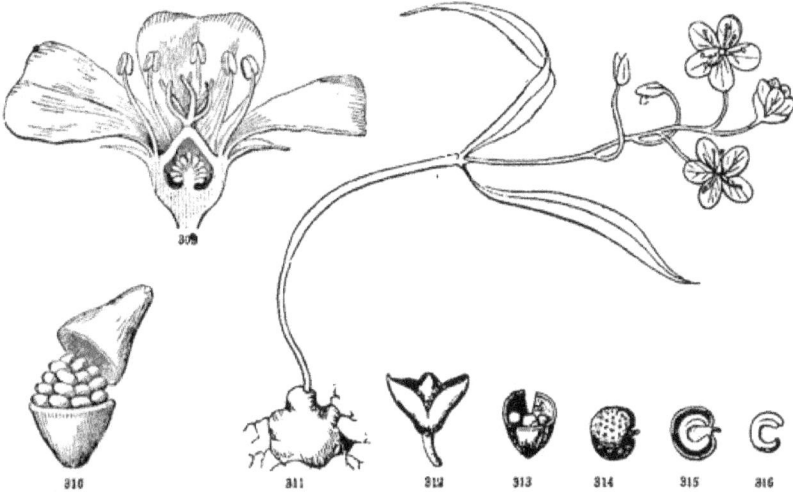

809. Half of a flower of the common Purslane, divided lengthwise and magnified. 810. Pod of the same, opening by a lid.
311. Claytonia or Spring-Beauty. 312 Its 2-cleft calyx and pod. 313. Ripe pod cut across, and splitting into three
valves. 314. Seed, more magnified. 315. Same, cut through, to show the coiled embryo. 316. Embryo taken out.

17. MALLOW FAMILY. Order MALVACEÆ.

Distinguished by the numerous *monadelphous* stamens (i. e, united by their filaments into
a tube or *column*), with kidney-shaped one-celled anthers, and the five sepals or lobes of the
calyx applied edge to edge without overlapping (i. e. *valvate*) in the bud, and persistent.
Leaves almost always palmately-veined, alternate, with stipules. Petals united at the bottom with the tube of stamens. There is often a sort of outer calyx, below the true one,
called an *involucel*. All innocent plants, full of mucilage (it is extracted from the root of
Marsh-Mallow), and with a very tough fibrous inner bark. Flowers often handsome.

Anthers all at the top of the column of united filaments (Fig. 317).
  Involucel or outer calyx present. Cells of the fruit many in a ring, separating whole
        when ripe, one-seeded.
      Involucel 9-parted. Separated little pods marginless. Plant soft-downy: root pe-
          rennial, (*Althæa*) MARSH-MALLOW.
      Involucel about 6-parted. Separated pods with membranaceous margins. Plants
          tall, roughish: root biennial. Flowers large, (*Althæa*, § *A'lcea*) * HOLLYHOCK.
      Involucel 3–6-cleft. A flat plate covering the circle of pods, (*Lavatèra*) * LAVATERA.
      Involucel 3-leaved. Circle of pods naked, around a narrow axis, (*Malva*) MALLOW.

Involucel or outer calyx none.
  Flowers diœcious, small, white. Pods or cells one-seeded,   (*Napǽa*) GLADE-MALLOW.
  Flowers perfect. Cells of the pod 5 to 15.
    Seed only one in each cell. Flowers yellow or white,   (*Sida*) SIDA.
    Seeds 2 to 9 in each of the cells,   (*Abútilon*) INDIAN-MALLOW.
Anthers attached along the sides of the upper part of the slender column. Pod of 3 to 5
                  cells, and splitting into as many valves.
  Involucel of many thread-shaped leaves.
    Calyx splitting down one side when the flower opens. Pod long, (*Abelmóschus*) * OKRA.
    Calyx not splitting down one side. Pod short. Seeds naked,   (*Hibiscus*) HIBISCUS.
  Involucel of 3 heart-shaped toothed leaves. Seeds bearing wool, (*Gossýpium*) * COTTON.

317, Stamens of Mallow united in a tube (monadelphous). 318. An anther more magnified. 319. Flowers and leaf of Marsh-Mallow. 320. Its compound pistil magnified. 321. Pod of Hibiscus surrounded by the calyx and involucel. 322. The pod splitting into 5 valves.

Mallow. *Malva.*

Involucel or outer calyx 3-leaved. Petals notched at the upper and broader end. Styles many. Little pods or cells many in a ring around a narrow axis or column (the whole shaped like a cheese), when ripe falling away separately, each one-seeded. — Herbs; flowering all summer.

1. LOW MALLOW. Root very long ; stems spreading on the ground ; leaves round-kidney-shaped, long-stalked, scarcely lobed, crenate; flowers several in the axils, small, whitish. Very common weed in waste and cultivated ground.   *M. rotundifolia.*

2. HIGH M. Stem 3° high; leaves lobed; flowers large, rose-purple. Gardens.   *M sylvéstris.*

3. MUSK M. Stem 2° high; leaves 5-parted and the divisions cut into linear lobes (the smell faintly musky); flowers large, rose-color. Gardens.   *M. moschàta.*

4. CURLED M. Stem 4° to 6° high ; leaves round, toothed, much curled around the edge ; flowers small, white, sessile in the axils. Gardens.   *M. crispa.*

Hibiscus. *Hibiscus.*

Flowers large, with an involucel of many narrow bractlets, and a 5-cleft calyx, which does not open down one side. Stamens in a long and slender column. Stigmas 5. Pod short, 5-celled, splitting when ripe into 5 valves, many-seeded; the seeds smooth or hairy, not long-woolly. Showy herbs or shrubs: flowering in autumn.

1. SHRUBBY or ALTHÆA HIBISCUS. Shrub 5° to 10° high, smooth ; leaves wedge-ovate, toothed, 3-lobed ; flowers short-stalked, white, purple-red, &c. (single or double). Cultivated for orna-ment. *H. Syriacus.*

2. GREAT RED H. Herb 8° high from a perennial root, smooth; leaves deeply cleft into 5 lance-linear lobes; corolla red, 8' to 11' broad! S. and in gardens. *H. coccineus.*

3. HALBERD-LEAVED H. Herb 6° high from a perennial root, smooth; lower leaves 3-lobed, upper halberd-shaped; calyx bladdery after flowering; corolla flesh-colored, 3' long. *H. militaris.*

4. MARSH H. Herb 6° high from a perennial root; leaves soft-downy and whitish underneath, ovate, pointed, the lower 3-lobed; base of the flower-stalks and leafstalks often grown together; corolla 5' broad, white or rose-color with a crimson eye. Salt marshes, &c. *H. Moscheutos.*

5. BLADDER-KETMIA H. (or *Flower-of-an-Hour*). Herb 1° to 2° high from an annual root, somewhat hairy; lower leaves toothed, upper 3-parted, with narrow divisions; corolla greenish-yellow with a dark-brown eye, opening only in midday sunshine; calyx bladdery after flowering, enclosing the pod. Gardens, &c. *H. Trionum.*

18. LINDEN FAMILY. Order TILIACEÆ.

Has the tough and fibrous inner bark and the bland mucilage of the Mallow family. Its distinctions are shown in the only genus we have, viz. : —

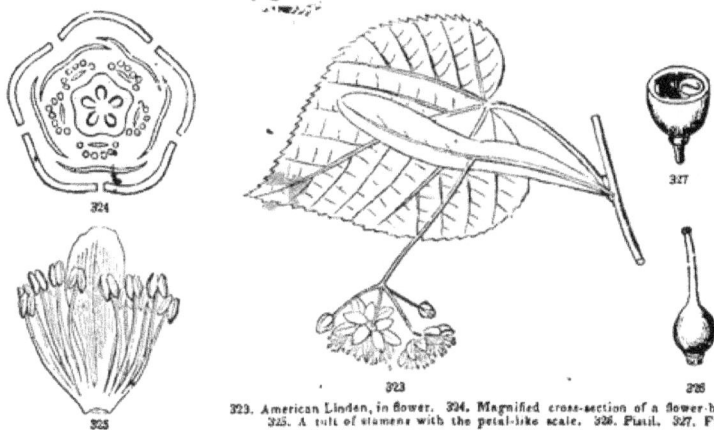

323. American Linden, in flower. 324. Magnified cross-section of a flower-bud. 325. A tuft of stamens with the petal-like scale. 326. Pistil. 327. Fruit cut in two.

Linden or Basswood. *Tilia.*

Sepals 5, thick, valvate (the margins edge to edge) in the bud, falling off after flowering. Petals 5, cream-color. Stamens very many, on the receptacle, in 5 clusters: anthers 2-celled. Pistil one: ovary 5-celled, with two ovules in each cell; in fruit woody, small, closed, mostly one-seeded. — Large, soft-wooded trees, with heart-shaped leaves, often oblique at the base. Flowers in a small cluster on a slender and hanging peduncle from the axil of a leaf, and united part way with a narrow leaf-like bract. (Also called *Lime-trees.*)

1. AMERICAN LINDEN or BASSWOOD.　Leaves green, smooth, or in some varieties downy underneath; a petal-like body in the middle of each of the 5 clusters of stamens.　　　　　*T. Americàna.*

2. EUROPEAN LINDEN.　Leaves smooth or nearly so; stamens hardly in clusters, no petal-like bodies with them.　Cultivated in cities, &c. as a shade-tree.　　　　　　*T. Europèa.*

19. CAMELLIA FAMILY.　Order CAMELLIACEÆ.

Shrubs or small trees, with alternate and simple leaves, not dotted; large and showy flowers, with a persistent calyx of 5 overlapping sepals, and very many stamens, their filaments united at the bottom with each other and with the base of the petals.　Anthers 2-celled.　Fruit a woody pod of 3 to 6 cells, containing few large seeds.　To this belongs the grateful TEA-PLANT of China, and the

CAMELLIA, of our green-houses,　　　　　　　　　　　*Caméllia Japónica.*
LOBLOLLY-BAY, of swamps in the Southern States,　　　*Gordònia Lasiánthus.*

20. ORANGE FAMILY.　Order AURANTIACEÆ.

Like the last, this family hardly claims a place here, being only house-plants, except far south.　Known by having 20 or more stamens in one row around a single pistil, and the leaves having a joint between the blade and the winged or margined footstalk: they (and the fragrant petals) are punctate with transparent dots, looking like holes when held between the eye and the light, which are little reservoirs of fragrant oil.　Fruit a berry with a thick rind.

ORANGE,　　　　　　　　　　　　　　　　　　*Citrus Auràntium.*
LEMON,　　　　　　　　　　　　　　　　　　*Citrus Limònium.*

21. FLAX FAMILY.　Order LINACEÆ.

Herbs with tough fibres in the inner bark, simple leaves, and oily seeds with a mucilaginous coat; consisting only of the Flax genus, which is known by the following marks: —

328. Common Flax.　　　329. Half of a flower, enlarged.　　　330. Pod, cut across.

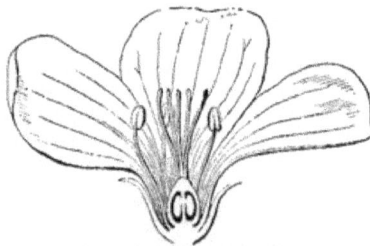

Flax.　*Linum.*

Sepals 5, overlapping, persistent.　Petals 5, on the receptacle.　Stamens 5, united with each other at the bottom.　Styles 5.　Pod 10-celled and splitting when ripe into 10 pieces with one seed in each.　Flowers opening only for one day.

1. COMMON FLAX. Root annual; leaves lance-shaped; flower blue. Cultivated. *L. usitatissimum.*

2. VIRGINIA FLAX. Root perennial; leaves oblong or lance-shaped; flowers very small, yellow. Dry woods. *L. Virginiànum.*

22. WOOD-SORREL FAMILY. Order OXALIDACEÆ.

Small herbs with sour juice, compound leaves of three leaflets, and flowers nearly as in the Flax family, but with 10 stamens, a 5-celled pod, and two or more seeds in each cell. One genus, viz.

Wood-Sorrel. *Óxalis.*

Sepals, petals, and styles 5. Stamens 10; filaments united (monadelphous) at the base. Pod thin, 5-lobed. Leaflets obcordate. Flowering in summer.

1. COMMON W. One-flowered scape and leaves rising from a scaly rootstock, hairy; petals large, white with reddish veins. N. in cold and moist woods. *O. Acetosélla.*

2. VIOLET W. Several-flowered scape and leaves, from a scaly bulb; petals violet. *O. violàcea.*

3. YELLOW W. Stems ascending, leafy; flowers 2 to 6 on one peduncle, small, yellow. *O. stricta.*

23. GERANIUM FAMILY. Order GERANIACEÆ.

Herbs or small shrubs, with scented leaves, having stipules; the lower ones opposite. Roots astringent. Sepals 5, overlapping. Petals 5. Stamens 10, but part of them in some cases without anthers : filaments commonly united at the bottom. Pistils 5 grown into one, that is, all united to a long beak of the receptacle (except the 5 stigmas); and when the fruit is ripe the styles split away from the beak and curl up or twist, carrying with them the five little one-seeded pods, as

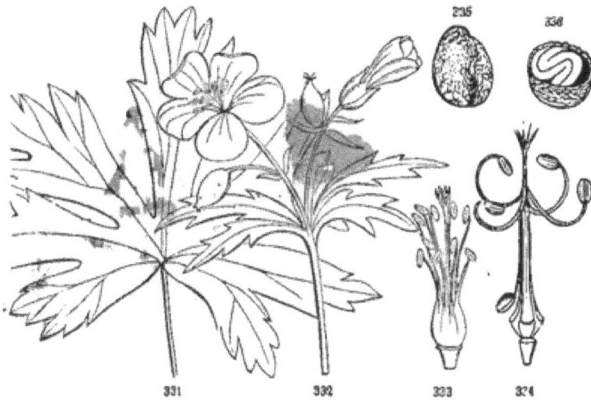

331. Leaf, and 332. Flowers of Wild Geranium. 333. Stamens and pistil. 334. Fruit bursting. 335. Seed. 336 Same, cut across.

shown in Fig. 334. — There are three genera, viz. GERANIUM or Cranesbill; ERODIUM, which differs in having only 5 stamens with anthers, and the fruit-bearing styles bearded inside ; and PELARGONIUM, which has the corolla more or less irregular, generally 7 stamens with anthers, &c. The latter are the House Geraniums, from the Cape of Good Hope, of several species and many varieties. We describe only the wild species of true

Geranium or Cranesbill. *Geranium.*

Petals all alike. All 10 stamens with anthers, every other one shorter. — Herbs.

1. SPOTTED G. Stem erect, from a perennial root; leaves 5-parted, also cut and toothed, often whitish-blotched; petals pale purple. Borders of woods; fl. in spring and summer. *G. maculàtum.*

2. CAROLINA G. Stems spreading from a biennial or annual root; leaves 5-parted, and cut into narrow lobes; flowers small; petals flesh-color, notched at the end. Waste places. . *G. Caroliniànum.*

3. HERB-ROBERT G. Stems spreading; leaves 3-divided, and the divisions twice pinnately cleft; flowers small, purple. Moist woods and ravines; fl. summer. *G. Robertiànum.*

24. INDIAN-CRESS FAMILY. Order TROPÆOLACEÆ.

Twining, climbing, or trailing herbs, with a watery juice of a sharp taste like Mustard, alternate leaves, and showy irregular flowers, as in

Indian-Cress (commonly called NASTURTIUM). *Tropæolum.*

Calyx projecting into a long hollow spur behind, petal-like, 5-cleft. Petals 5, of two sorts, two of them borne on the throat of the calyx, the 3 others with claws. Stamens 8, unequal. Fruit 3-lobed, separating into 3 thick and closed one-seeded pieces.

1. COMMON I. or NASTURTIUM. Very smooth; leaves roundish, shield-shaped; flowers large; petals orange-yellow, the claws of 3 of them fringed. Cult. very common. *T. majus.*

2. CANARY-BIRD I. Climbing high; leaves deeply lobed; petals pale yellow, cut-fringed. Cult. *T. peregrinum.*

25. BALSAM FAMILY. Order BALSAMINACEÆ.

Tender annuals, with a bland watery juice and very irregular flowers; such as those of the principal genus,

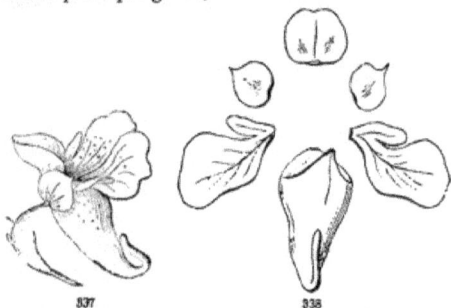

337. Flower of No. 2.   338. Calyx and corolla displayed.

Balsam (or JEWEL-WEED). *Impàtiens.*

Calyx and corolla colored alike and difficult to distinguish, in all of 6 pieces, the largest one extended backward into a large and deep sac ending in a little spur; and the two innermost unequally 2-lobed. Stamens on the receptacle, 5, very short, united over the pistil. This forms a thick-walled pod, which when ripe suddenly bursts with considerable force, or falls into 5 coiling pieces at the touch, scattering the rather large seeds. — Leaves simple, alternate. Flowers showy, produced all summer.

1. GARDEN BALSAM. Flowers very showy, white, red, or pink, often double, clustered in the axils of the crowded lance-shaped leaves. Garden annual. *I. Balsàmina.*

2. PALE JEWEL-WEED. Flowers pale-yellow, sparingly spotted, the hanging sac broader than long; leaves ovate or oblong. Common in rich and shady or wet soil. *I. pàllida.*

3. SPOTTED JEWEL-WEED. Flowers orange, spotted with reddish-brown; sac longer than broad. *I. fulva.*

26. RUE FAMILY. Order RUTACEÆ.

Strong-scented, sharp-tasted, and bitter-acrid plants, the leaves dotted with transparent dots like punctures (which are filled with volatile oil) ; the stamens on the receptacle, as many or twice as many as the petals.

Herbs, very strong-scented, with perfect flowers. Stamens 8 or 10.

Leaves decompound. Flowers yellow: petals concave. Pod roundish, *(Ruta)* * RUE.

Leaves pinnate. Flowers white or purple, large: petals slender: stamens long. Pods 5, flattened, slightly united, *(Dictamnus)* * FRAXINELLA.

Shrubs or trees. Stamens 4 or 5, only as many as the petals.

Flowers diœcious. Pistils 2 to 5, making fleshy pods with one or two black seeds. Leaves pinnate. Stems prickly, *(Zanthóxylum)* PRICKLY-ASH.

Flowers polygamous. Pistil 1, making a 2-celled, 2-seeded key, winged all round. Leaflets 3. Stems not prickly, *(Ptèlea)* HOP-TREE.

27. SUMACH FAMILY. Order ANACARDIACEÆ.

Trees or shrubs with a milky or a resinous-acrid juice (in some cases poisonous), and alternate leaves : — of which we have only the genus

Sumach. *Rhus.*

Flowers small, greenish-white or yellowish. Sepals, petals, and stamens 5; the latter borne on an enlargement of the receptacle which fills the bottom of the calyx. Styles or stigmas 3, on a one-celled ovary; which makes a one-seeded little stone-fruit with a thin flesh. Fl. summer. Nos. 4 and 5 are poisonous to most people when touched.

1. STAGHORN SUMACH. Small tree ; branches and stalks velvety-hairy; leaves pinnate, pale beneath; flowers and crimson-hairy sour fruit very many, in a great crowded panicle. *R. typhina.*

2. SMOOTH S. Shrub; branches and stalks very smooth, pale: otherwise like the last. *R. glabra.*

8. DWARF S. Shrub 1° to 4° high; branches and stalks downy; leaves pinnate, with the stalk wing-margined between the shining leaflets; fruits red and hairy. *R. copallina.*

4. POISON S. or DOGWOOD. Shrub smooth; leaves pinnate; leaflets 7 to 13, entire; panicles slender in the axils, fruit smooth. *Poisonous* to most people. Swamps. *R. venenàta.*

5. POISON IVY. Smooth; stems climbing by rootlets; leaflets 3, large, ovate, either entire, notched, or lobed, variable on the same stem. Poisonous like the last. *R. Toxicodéndron.*

6. VENETIAN S. or SMOKE-TREE. Shrub, with simple oval or obovate leaves ; branches of the panicle lengthening after flowering, and feathered with long hairs, making large light bunches. Cult. *R. Cótinus.*

28. GRAPE FAMILY. Order VITACEÆ.

Shrubby plants with a watery and sour juice, climbing by tendrils; known by having a minute calyx with scarcely any lobes, the petals valvate (edge to edge) in the bud and falling off very early, and the stamens (5 or 4) one before each petal ! — Only two genera.

Grape. *Vitis.*

Petals 5, cohering slightly at the top while they separate at the base, and generally thrown off without expanding. Berry with 4 bony seeds. Leaves lobed. Flowers polygamous in the wild species, and having the fragrance of Mignonette.

1. EUROPEAN GRAPE.   Flowers all perfect; leaves deeply and sharply lobed.   Cult. in several varieties, viz. Sweetwater Grape, Black Hamburg, &c.   *V. vinifera.*

2. NORTHERN FOX-GRAPE.   Leaves very woolly when young, remaining rusty-woolly beneath; berries large, purple or amber-colored. — Improved varieties of this, without the foxy taste and the
tough pulp, are the Isabella and the Catawba Grapes.   *V. Labrúsca.*

3. SUMMER GRAPE.   Leaves with loose cobwebby down underneath, smoothish when old ; panicles of
fertile flowers very long and slender; berries small, ripe with first frost.   *V. æstivàlis.*

4. FROST GRAPE.   Leaves thin, heart-shaped, never woolly, not shining, sharply and coarsely toothed,
little or not at all lobed ; panicles loose ; berries blue or black with a bloom, sour, ripening late.
Common along river-banks, &c.   *V. cordifòlia.*

5. MUSCADINE or SOUTHERN FOX-GRAPE.   Bark of the stem close, not thrown off in loose strips, as
in the others; leaves round-heart-shaped, shining, not downy, very coarsely toothed; panicles small,
with crowded flowers; berry large, musky, with a very thick and tough skin.   A variety is the
Scuppernong Grape.   Common S.   *V. vulpina.*

Virginia-Creeper. *Ampelópsis.*

Petals 5, thick, opening before
they fall. Leaves palmate with 5
leaflets (Fig. 74). Berries small,
blackish. A very common tall-
climbing vine, wild and cultivated.   *A. quinquefòlia.*

340. Flower opening. 341. Same, with the   339. Twig of Grape-vine.
                     petals fallen.

29. BUCKTHORN FAMILY.   Order RHAMNACEÆ.

Woody plants, with simple alternate leaves, known by having the stamens as many as the
small petals (4 or 5) and one before each of them,
both inserted on the calyx or on a fleshy cup which
lines the tube of the calyx ; the lobes of the latter
valvate, i. e. edge to edge in the bud.   Fruit of 2 to
5 cells, and one large seed in each.

342. Flowers of a Buckthorn. 343. Same, cut through lengthwise.

342   343

Calyx free from the ovary, greenish. Petals shorter than the calyx, or none,   (*Rhamnus*) BUCKTHORN.
Calyx below adherent to the ovary, its lobes petal-like (white in our species) and
bent inwards, shorter than the stamens and long-clawed petals,   (*Ceanòthus*) NEW-JERSEY TEA.

30. STAFF-TREE FAMILY. Order CELASTRACEÆ.

Woody plants, with simple alternate or opposite leaves; the divisions of the calyx and the petals both overlapping in the bud; the stamens as many as the petals (4 or 5) and alternate with them, inserted on a thick expansion of the receptacle (disk) which fills the bottom of the calyx. Pod colored, of 2 to 5 mostly one-seeded cells, showy when ripe in autumn, especially when they open and display the seeds enveloped in a pulpy scarlet aril.

Flowers polygamous or nearly diœcious, white, in racemes: disk cup-shaped: style long.
    Pod globular, orange-yellow. Leaves alternate. Our only species is a twin-
      ing shrub, sometimes called BITTERSWEET,   (*Celástrus*) WAXWORK.
Flowers perfect, flat, dull green or dark purple, in axillary racemes: disk flat, covering
    the ovary, and bearing 4 or 5 very short stamens, the short style just rising
      through it. Pods red, lobed. Shrubs: leaves opposite,   (*Euónymus*) BURNING-BUSH
Pods smooth, strongly lobed,   or SPINDLE-TREE.
Pods roundish, rough,   (*Euónymus*) STRAWBERRY-BUSH.

31. SOAPBERRY FAMILY. Order SAPINDACEÆ.

The proper Soapberry family belongs mostly to warmer climates; but we have shrubs and trees belonging to three of its subfamilies:

347   344   345   345

I. BLADDERNUT SUBFAMILY. Flowers regular and perfect. Stamens 5, as many as the petals, and alternate with them. Seeds bony. Leaves opposite, pinnate or with 3 leaflets, having stipules, and also little stipules (*stipels*) to the leaflets.

Shrub: flowers white in racemes. Fruit of 3 bladdery
    pods united.   (*Staphylèa*) BLADDERNUT.

348

II. HORSECHESTNUT SUBFAMILY. Flowers polygamous, some of them having no good pistil, mostly irregular and unsymmetrical. Calyx bell-shaped or tubular, 5-toothed. Petals 4 or 5, with claws, on the receptacle. Stamens generally 7, long. Style one. Ovary 3-celled, with a pair of ovules in each cell, only one or two ripening in the fruit; which becomes a leathery 3-valved pod. Seeds very large, like chestnuts. Fine ornamental trees, with opposite palmate leaves, and flowers in thick panicles.

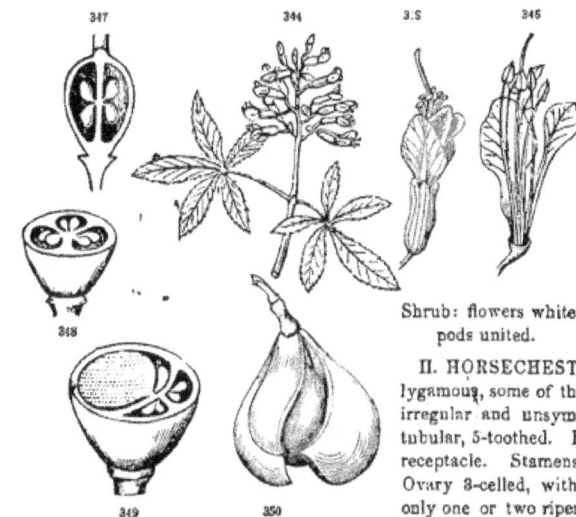

349   350

344. Red Buckeye, reduced in size. 345. Flower 346. Same, with calyx and two petals taken away. 347. Magnified ovary, divided lengthwise. 348. Same, divided crosswise, showing the two ovules in each cel.. 349. Same, partly grown, only one seed growing. 350. Ripe pod bursting.

Petals 5, spreading; stamens declined: fruit prickly. Leaflets 7,   (*Æsculus*) *HORSECHESTNUT.
Petals 4, unlike, with long claws in the calyx. Leaflets generally 5,   (*Æsculus*, § *Pàvia*) BUCKEYE.

III. MAPLE SUBFAMILY. Flowers generally polygamous or dirœcious, regular. Petals often none, but the calyx sometimes petal-like. Stamens 4 to 12. Styles 2, united below. Fruit a pair of keys united at the bottom (Fig. 208). Leaves opposite.

Flowers diœcious, small and greenish: petals none: stamens 4 or 5. Leaves pinnate,
    with 3 to 5 veiny leaflets: twigs green,   (*Negúndo*) NEGUNDO.
Flowers polygamous or perfect. Leaves simple, palmately lobed,   (*Acer*) MAPLE.

Buckeye. *Æsculus, § Pàvia.*

All wild species at the West and South: also cultivated for ornament: flowering in late spring or summer.

1. FETID or OHIO BUCKEYE. Petals small, erect, pale yellow, shorter than the curved stamens; young fruit prickly like Horsechestnut; a tree. River-banks, W.   *Æ. glabra.*

2. SWEET BUCKEYE. Petals yellow or reddish, erect, enclosing the stamens; fruit smooth. *Æ. flava.*

3. RED BUCKEYE. Petals red, also the tubular calyx: otherwise like the last. Shrub. *Æ. Pàvia.*

4. SMALL-FLOWERED B. Leaflets stalked; petals white, rather spreading; stamens very long; fruit smooth; seed eatable, not bitter, as are the others; flowers in a long raceme-like panicle. Shrub. S. & cult.   *A. parviflòra.*

Maple. *Acer.*

  * Flowers in terminal racemes, with petals, greenish, in late spring: stamens 6 to 8.

1. STRIPED MAPLE. Bark green, with darker stripes; leaves large, with 3 short and taper-pointed lobes; racemes hanging. Small tree in cool woods; common, N. *A. Pennsylvánicum.*

2. MOUNTAIN M. Bark gray; leaves 3-lobed; racemes erect; flowers small. Shrub, N. *A. spicàtum.*

3. SYCAMORE M. An imported shade-tree, with large strongly 5-lobed leaves, and large hanging racemes, flowering soon after the leaves appear. *A. Pseudo-Plátanus.*

  * * Flowers in loose clusters, yellowish-green, appearing with the leaves, in spring.

4. NORWAY M. An imported shade-tree, with leaves resembling Sugar Maple, but brighter green on both sides, rounder, and with some long pointed teeth; flowers in an erect terminal corymb, with petals; wings of the fruit very large, diverging. *A. platanoìdes.*

5. SUGAR or ROCK M. Leaves with 3 or mostly 5 long-pointed lobes, their edges entire except a few coarse wavy teeth; flowers hanging on very slender hairy stalks, without petals; fruit with rather small wings, ripe in autumn. Tall tree; in rich woods, and commonly planted for shade. *A. sacchárinum.*

  * * * Flowers in early spring, considerably earlier than the leaves, on short pedicels, in small umbel-like clusters from lateral leafless buds: stamens generally 5: fruit ripe and falling in early summer.

6. WHITE or SILVER M. Leaves very deeply 5-lobed, cut and toothed, white beneath; flowers greenish-yellow, short-stalked, without petals ; fruit woolly when young, with very large and smooth diverging wings. Tree common on river-banks, and planted for shade. *A. dasycárpum.*

7. RED or SOFT M. Leaves whitish beneath, with 3 or 5 short lobes, toothed; flowers on very short stalks which lengthen in fruit, with linear-oblong petals, red or sometimes yellowish; wings of the fruit small, reddish. Wet places: a common tree. *A. rubrum.*

32. PULSE FAMILY. Order LEGUMINOSÆ.

A large family, distinguished by the peculiar irregular corolla called *papilionaceous* (i. e. butterfly-shaped), and for having the kind of pod called a *legume* for its fruit. Leaves alternate, often compound, with stipules. Stamens generally 10, inserted on the calyx. Pistil one, simple. The papilionaceous corolla, which is familiar in the Pea-blossom and the like, consists of 5 irregular petals; viz. an upper one, generally largest and outside in the bud, called the *standard;* two side petals, called *wings,* and two lower ones put together and commonly a little joined, forming a kind of pouch which encloses the stamens and style, and which, being shaped somewhat like the prow of an ancient vessel, is named the *keel.* A few flowers in the family are almost regular, or not papilionaceous. In one case (to be mentioned in its place) all but one petal is wanting. Another set have perfectly regular blossoms; but are known by the pod and leaves. The legume is of every variety of shape and size. The whole kernel of the seed is an embryo, with thick cotyledons, as is familiar in the Bean and Pea (Fig. 32, 42). We give the principal sorts.

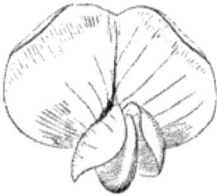

351. Papilionaceous corolla of Locust.

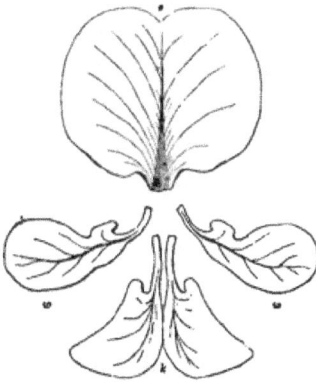

352. Its petals displayed : *s*, standard ; *w, w,* wings ; *k,* the keel laid open.

353. Legume of Pea, open.

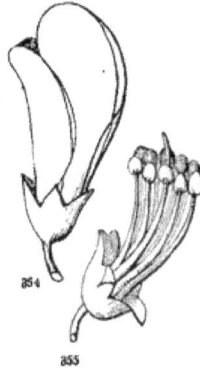

354. Flower of False Indigo No. 2. 355. Same, with the petals removed.
356. Flower of Amorpha, enlarged. 357. Stamens and pistil of the same.

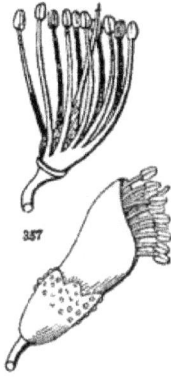

I. TRUE PULSE FAMILY. Corolla really papilionaceous, and the standard outside, wrapped around the other petals in the bud (in Amorpha, Fig. 356, only the standard is present). Leaves either simple or only once compound.

Stamens, 10, united by their filaments, either all into a closed tube (*monadelphous*, Fig. 187), or 9 in a tube split down on one side, and the 10th separate or nearly so (*diadelphous*, Fig. 188).

Shrubs or trees, not twining nor climbing.
  Flowers white or rose-colored, in hanging racemes. Leaves odd-pinnate, (*Robinia*) LOCUST-TREE.
  Flowers yellow, in small racemes. Pod bladdery. Leaves odd-pinnate, (*Colútea*) *BLADDER-SENNA.
  Flowers yellow, in hanging racemes. Pod narrow. Leaflets 3,   (*Cýtisus*) * LABURNUM.
Shrubs, with long twining stems. Flowers blue-purple in racemes,   (*Wistária*) WISTARIA.
Herbs.
  Stems not twining, climbing, nor with any trace of tendrils.
    Leaves simple: stipules winging the stem below the leaf. Flowers yellow. Pod
        inflated, many-seeded,   (*Crotalària*) RATTLEBOX.
    Leaves of 5 to 15 palmate leaflets. Flowers in a long raceme,   (*Lupinus*) LUPINE.
    Leaves abruptly pinnate, of 4 leaflets. Pod formed underground,   (*A'rachis*) * PEANUT.
    Leaves odd-pinnate, of several or many leaflets.
      Leaflets serrate. Flowers single, white. Pod inflated, 2-seeded, (*Cicer*) * CHICK-PEA.
      Leaflets entire. Flowers in a raceme or spike.
        Corolla broad. Pod flat, narrow, several-seeded, (*Tephròsia*) HOARY-PEA.
        Corolla narrow. Pod inflated or turgid, often 2-celled, (*Astrágalus*) ASTRAGAL.
    Leaves of 3 (or rarely 5) leaflets. Pods like akenes or burs.
      Stipules cohering with the base of the leafstalk.
        Flowers in heads. Pod thin and small, in the persistent calyx, (*Trifòlium*) CLOVER.
        Flowers in spikes or racemes.
          Leaflets not dotted, but generally serrate.
            Pods wrinkled, like akenes, coriaceous,   (*Melilòtus*) MELILOT.
            Pods curved, or else coiled up in various ways,   (*Medicàgo*) MEDICK.
          Leaflets, &c. sprinkled with dark dots, entire.   (*Psorálea*) PSORALEA.
      Stipules not at all united with the leafstalk.
        Pods very small and flat, closed, one-seeded,   (*Lespedèza*) BUSH-CLOVER.
        Pods separating into several-seeded flat roughish joints, (*Desmòdium*) TICK-TREFOIL.
Stems climbing or disposed to climb : leaves pinnate with a tendril at the end.
  Calyx with 5 leafy lobes. Seeds globular. Leaflets few,   (*Pisum*) * PEA.
  Calyx-lobes or teeth not leafy.
    Tendril conspicuous. Style hairy along the inner side, (*Làthyrus*) EVERLASTING-PEA.
    Tendril conspicuous. Style hairy round the tip,   (*Vícia*) VETCH.
    Tendril hardly any. Seed oblong, fixed by one end,   (*Faba*) * HORSE-BEAN.
Stems twining more or less: no tendrils to the leaves.
  Keel of the corolla coiled into a ring or spiral.
    Leaflets 3, with stipels,   (*Phasèolus*) BEAN.
    Leaflets 5 or 7. Flowers brown-purple. Tubers underground,   (*A'pios*) GROUNDNUT.
  Keel not coiled or twisted. Leaflets 3, with stipels.
    Calyx 4-cleft, the lobes acute,   (*Galáctia*) MILK-PEA.
    Calyx 4-toothed. Pods both above and under ground, (*Amphicarpèa*) HOG-PEANUT.
    Calyx tubular, 5-toothed. Flower large, purple-blue,   (*Clitòria*) BUTTERFLY-PEA.
Stamens 10, separate, except at the very base. Petal only one! (Fig. 356.) Shrubs:
  leaves pinnate: flowers small, violet-purple, in a spike or raceme, (*Amórpha*) AMORPHA.

Stamens 10, separate (Fig. 355). Petals 5, pea-like. Pod inflated, (*Baptisia*) FALSE-INDIGO.

II. BRASILETTO SUBFAMILY. Corolla sometimes papilionaceous or nearly so, but then with the standard within the other petals, generally more or less irregular; the petals overlapping one another in the bud. Stamens 10 or fewer, separate.

Trees, with simple round-heart-shaped leaves, but appearing rather later than the papilio-
    naceous purple-red flowers, (*Cercis*) RED-BUD.
Herbs, with abruptly pinnate leaves and yellow flowers, not papilionaceous, (*Cassia*) SENNA.
Trees, with the leaves, or some of them, more than once compound. Flowers diœcious
    or polygamous, not at all papilionaceous.
    Stamens 10, and petals 5, on the top of the funnel-shaped tube of the calyx. Pods
      broad and hard. Leaves very large, twice-pinnate, (*Gymnocladus*) KENTUCKY COFFEE-TREE.
    Stamens and petals 3 to 5, on the bottom of an open calyx. Pods long and flat, hav-
      ing a sweet juice or pulp inside. Leaves, some of them once pinnate, others twice
      pinnate. Tree with compound thorns, (*Gleditschia*) HONEY-LOCUST.

III. MIMOSA SUBFAMILY. Flowers very small, in heads or spikes, regular: petals edge to edge in the bud, and sometimes united below. Leaves generally twice or thrice pinnate.

Stamens very many and long, yellow or yellowish. (Cult. in greenhouses: some species
    are wild far South), * ACACIA.
Stamens 5. Petals separate, whitish. Pod smooth, (*Desmanthus*) DESMANTHUS.
Stamens 4 or 5. Petals united into a cup, rose-color. Pod bristly, flat, breaking up into
    joints. Leaves closing suddenly when touched, (*Mimosa*) * SENSITIVE-PLANT.
Stamens 10 or 12. Petals united into a cup, rose-color. Pod narrow, rough-prickly.
    Leaves rather sensitive. S, (*Schrankia*) SENSITIVE-BRIER.

Locust-tree. *Robinia.*

Flowers showy, in hanging axillary racemes. Stamens diadelphous. Pod flat, several-seeded. Leaves odd-pinnate. — Trees, wild in the Southern, cult. in the Northern States. Fl. in early summer.

1. COMMON LOCUST-TREE. Tree with a pair of spines for stipules; flowers white, in slender racemes, sweet-scented; pod smooth. *R. Pseudacàcia.*

2. CLAMMY L. Tree with clammy twigs; racemes thick; calyx purplish; pod rough. *R. viscòsa.*

3. BRISTLY L. or ROSE-ACACIA. Shrub, with bristly stalks and twigs; flowers large, rose-colored. *R. hispida.*

Clover (or TREFOIL). *Trifòlium.*

Flowers many in a head. Calyx persistent, its teeth very slender. Corolla withering away or persistent after flowering; the petals grown together more or less into a tube below, and the diadelphous stamens united with it. Pod generally shorter than the calyx, thin, only one- or few-seeded. Low herbs: leaves with 3 leaflets, the stipules adhering to the base of the footstalk (Fig. 136).

1. RED CLOVER. Leaflets obovate or oval, with a pale spot on the upper side; flowers rose-red, in a dense head with leaves underneath it. Fields, cultivated. *T. praténse.*

2. BUFFALO C. Leaflets obovate, toothed : flowers rose-colored, pedicelled, in an umbel-like long-stalked head. Prairies, &c., W. & S. *T. reflèxum.*

3. WHITE C. Low, smooth, creeping; leaflets obcordate or notched; flowers white, in a loose umbel-like head, raised on a long stalk. Fields, &c., everywhere. *T. repens.*

4. **RABBIT-FOOT** C. Silky, low, erect, and branching; root annual; leaflets narrow; flowers whitish, in dense and soft-silky oblong heads. Common in poor dry land.   *T. arvénse.*

5. **YELLOW** C. Low, annual, smoothish; corolla yellow, turning brownish. Waste grounds.

T. agrárium-

Melilot (or SWEET-CLOVER). *Melilótus.*

Flowers in a raceme or spike, small. Corolla falling after flowering. Pod roundish and small, like an akene, hardly opening, containing only one or two seeds. — Annuals or biennials, with sweet-scented foliage; leaflets three, toothed. Growing in gardens and around houses.

1. **YELLOW MELILOT.** Leaflets obovate or oblong, obtuse; corolla light yellow.   *M. officinàlis.*
2. **WHITE M.** Leaflets as if cut off square at the end; corolla white.   *M. alba.*

Medick. *Medicàgo.*

Flowers like those of Melilot, either few or many in a cluster. Pod curved or coiled, either kidney-shaped or rolled up spirally in various ways. Leaves of 3 leaflets.

1. **LUCERNE,** or **PURPLE MEDICK.** Stems upright from a deep perennial root; leaflets obovate-oblong ; flowers purple in short racemes ; pods spiral. Cultivated for green fodder.   *M. sativa.*
2. **BLACK M.** Stems reclining ; leaflets wedge-obovate; flowers yellow, in short spikes; pods curved (Fig. 358), wrinkled, turning blackish. Waste grounds.   *M. lupulìna.*
3. **SNAIL M.,** with 2-flowered peduncles, is sometimes cultivated in gardens, on account of its singular pods coiled like a shell (Fig. 359).   *M. scutellàta.*

Everlasting-Pea or Vetchling. *Láthyrus.*

Lobes or teeth of the calyx not leafy. Style flattish. Otherwise the flowers nearly the same as in the true Pea.

* Garden species, cultivated for ornament; with winged stems and only one pair of leaflets.

1. **SWEET PEA.** Root annual; flowers 2 or 3 on a long peduncle, sweet-scented   *L. odorátus.*
2. **GARDEN EVERLASTING-PEA.** Root perennial; flowers many, pink or purple.   *L. latifòlius.*

* * Wild species, with perennial roots and more than one pair of leaflets.

3. **MARSH E.** Stems lightly winged or margined; leaflets 2 to 4 pairs, lance-linear or lance-oblong; stipules lance-shaped; flowers 2 to 5, purple. Moist ground, N.   *L. palústris.*
4. **PALE E.** Leaflets 3 or 4 pairs, ovate, pale; stipules rather large, half heart-shaped; flowers 7 to 10, cream-color. Banks and thickets, W. & N.   *L. ochroleùcus.*
5. **VEINY E.** Leaflets 5 to 7 pairs, oblong or ovate; stipules very small; flowers many on the peduncle, purple. Shady banks, S. & W.   *L. venòsus.*
6. **BEACH PEA.** Leaflets 4 to 6 pairs, oval or obovate; stipules large and leafy; flowers 6 to 10 on the peduncle, purple. Shore of the sea, N. and of the Great Lakes.   *L. maritìmus.*

Vetch or Tare. *Vicia.*

Like the last, but with small and usually more numerous leaflets; and the thread-shaped style hairy round the end or down the outer side.

* Perennials, all wild species: flowers small, in a raceme on a long peduncle.

1. **TUFTED V.** Downy ; leaflets many, lance-oblong, strongly mucronate ; flowers crowded, bent down in the spike, blue, turning purple, summer. Thickets, N.   *V. Cracca-*

2. Carolina V. Smooth; leaflets 8 to 12, oblong; flowers many, whitish, tipped with blue, rather scattered on the peduncle, in spring. Banks, &c., common. *V. Caroliniána.*

3. American V. Smooth; leaflets 10 to 14, oval or oblong, very veiny; flowers 4 to 8 on the peduncle, purplish or bluish, in summer. N. *V. Americúna.*

* * Annual: flowers large, one or two together, sessile in the axils of the leaves.

4. Common Tare. Leaflets 10 to 14, narrow; flowers violet-purple. Cultivated fields, *V. satíva.*

Bean. *Phaséolus.*

Keel of the corolla (with the included stamens and style) twisted or coiled, so as to form a ring, or one or more turns of a spiral coil. Stamens diadelphous. Pod flat or flattish, several-seeded. Seeds flattish. Plants twining more or less, in one cultivated variety short and erect. Leaves of three leaflets, the end leaflet some way above the other two (i. e. pinnate of 3 leaflets): and they have *stipels* or little stipules to the leaflets. Fl. summer.

* Wild species: mostly found South and West.

1. Perennial Bean. Climbing high; leaflets round-ovate, pointed; flowers in long panicled racemes, purple; pods curved. Wooded banks, &c. *P. perénnis.*

2. Trailing Bean. Annual, spreading on the ground ; leaflets 3-lobed or angled ; flowers few, crowded at the end of a long erect peduncle, purplish; pods narrow, straight. Sandy places. *P. diversifólius.*

* * Cultivated Beans.

3. Common or Kidney Bean. Known by its straight pods, pointed by the hardened lower part of the style, and the thick rather kidney-shaped seeds. The Dwarf or Bush Bean is a low and small variety which does not twine. The Scarlet Runner is a free climbing variety, generally red-flowered. *P. vulgáris.*

4. Lima Bean. Known by its broad and flat, curved or scymitar-shaped pods, with few and large flat seeds. The Civet Bean is a small variety of it. *P. lunátus.*

False-Indigo. *Baptisia.*

Flowers generally in racemes. Standard erect, with the sides rolled back: keel-petals nearly separate and straight, like the wings. Stamens 10, separate! Pod stalked in the calyx, bladdery, but rather thick-walled, pointed, containing many small seeds. — Perennial herbs, erect and branched, with palmate leaves of 3 leaflets. — The commonest are the following: —

1. Yellow False-Indigo. Glaucous, bushy-branched; leaves almost sessile; leaflets small, wedge-obovate; flowers few at the ends of the panicled branchlets, yellow, produced all summer. Dry grounds, common. *B. tinctória.*

2. Blue F. Tall and stout; stipules lance-shaped, as long as the petiole; leaflets wedge-oblong; flowers many, large, blue, in a long raceme, in spring or early summer. (Fig. 354, 355.) Rich soil; common W. & S. and also cultivated in gardens. *B. austrális.*

Senna. *Cássia.*

Calyx of 5 sepals. Petals 5, spreading, not papilionaceous, but a little irregular. Stamens 10, but those on one side of the blossom commonly shorter, or without anthers; the anthers open at the top by two chinks or holes. Pods many-seeded. — Leaves simply and abruptly pinnate. The common species are herbs, with yellow flowers, in summer.

1. MARYLAND SENNA. Root perennial; stems 3° or 4° high; leaflets 6 to 9 pairs, lance-oblong, 1' or more long, used for medicine instead of the imported *senna*. Rich soil.   *C. Marilándica.*

2. PARTRIDGE-PEA S. Annual, low, spreading; leaflets 10 to 15 pairs, linear-oblong, ½' long; flowers large and showy; anthers 10, six of them purple. Sandy fields.   *C. Chamæcrista.*

3. SENSITIVE S. Flowers small, short-stalked; anthers only 5: otherwise like the last.   *C. nictitans.*

33. ROSE FAMILY.   Order ROSACEÆ.

A large and most important family of plants, distinguished by having alternate leaves with stipules, and regular flowers; their generally 5 petals (sometimes wanting) and sta-

mens (generally numerous, at least above 10) inserted on the persistent calyx. The seeds are few and their whole kernel is embryo, as is seen in an almond (Fig. 36), Apple-seed, or Cherry-seed (Fig. 38), &c. The family furnishes some of our most esteemed fruits: all the plants are innocent, except the strong-scented foliage and bark, in the Almond sub-family. For figures illustrating this family, see those of Cherry-blossom (Fig. 193), Hawthorn-blossom (Fig. 194), the fruit of Apple and Quince, (Fig. 200 and 201), Peach (Fig. 202), Rose and Strawberry (Fig. 220 –222), and the annexed figures.

360. Section of a Rose-bud.   361. American Crab-Apple.

I. ALMOND SUBFAMILY. Pistil only one, free from the calyx, becoming a stone-fruit. — Trees or shrubs with simple leaves ; the bruised bark and foliage with a peculiar aromatic scent and flavor. — The plants of this division are all ranked under two great genera (*Amygdalus* and *Prunus*), but under several subgenera, here adopted for the convenience of the common names.

Calyx with a rather deep cup.   Petals rose or red-purple.   Stone of the fruit rough.
    Flesh of the fruit becoming a dry husk. We have the dwarf Flowering-Almond in
         gardens, with double flowers. It does not form fruit here,   (*Amygdalus*) *ALMOND.*
    Flesh pulpy: surface downy (or in NECTARINE smooth),   (*Pérsica*) *PEACH.*
Calyx with a short and broad cup.   Petals white.   Stone of the fruit smooth, and
    Flattened, with grooved edges: skin of the fruit downy,   (*Armeniaca*) *APRICOT.*
    Flat or flattish, generally edged: fruit smooth, with a bloom,   (*Prunus*) PLUM.
    Roundish or globular: fruit smaller, smooth, without a bloom,   (*Cerasus*) CHERRY.

II. ROSE SUBFAMILY. Pistils few or many (rarely only one), separate from each other and *free* from the persistent calyx, but sometimes (as in the Rose, Fig. 360) enclosed and concealed in its tube. Stipules generally united with the bottom of the leafstalk on each side.

Pistils generally 5, making few-seeded pods.

Petals broad: calyx open, 5-cleft. Shrubs or herbs,   (*Spiræa*) MEADOW-SWEET.

Petals lance-shaped: calyx narrow, 5-toothed. Herbs,   (*Gillenia*) INDIAN-PHYSIC.

Pistils only one or two, making akenes, enclosed in the narrow-mouthed tube of the calyx.

Petals 5, yellow: stamens 12 or more: calyx bur-like,   (*Agrimònia*) AGRIMONY.

Petals none; but the 4 spreading lobes of the smooth calyx petal-like.

    Flowers perfect, in a spike: stamens 4, long (white),   (*Sanguisórba*) BURNET.

    Flowers monœcious, in a head: stamens many,   (*Potèrium*) * SALAD-BURNET.

Pistils 3 to 10, making akenes: stamens many. (Stemless herbs.)

Petals 5, yellow. Leaves of 3 leaflets,   (*Waldsteinia*) BARREN-STRAWBERRY.

Petals 5, white. Leaves simple, rounded-heart-shaped,   (*Dalibárda*) DALIBARDA.

Pistils many, making akenes, or in Bramble berry-like in fruit.

Calyx open, with 5 additional outer lobes (making 10) or 5 accessory teeth.

    Akenes tipped with a long feathery or hooked or twisted tail (style),   (*Geum*) AVENS.

    Akenes seed-like; the short style falling off.

      Receptacle of the fruit dry and small,   (*Potentilla*) CINQUEFOIL.

      Receptacle of the fruit becoming very large and pulpy,   (*Fragária*) STRAWBERRY.

Calyx open, flat, 5-lobed. Ovaries in a head, becoming berry-like,   (*Rubus*) BRAMBLE.

Calyx with an urn-shaped or globular closed tube and 5 lobes,   (*Rosa*) ROSE.

III. PEAR SUBFAMILY. Pistils 2 to 5, their styles more or less separate, their ovaries united with each other and with the thick tube of the calyx which encloses them and makes a fleshy fruit (*pome*). Stipules free from the leafstalk. Trees or shrubs.

Cells of the fruit containing only one or two seeds.   [or SHADBUSH.

Petals long and narrow. Fruit berry-like, its cells becoming 10,   (*Amelánchier*) JUNE-BERRY

Petals broad or rounded.

    Fruit drupe-like, containing 2 to 5 stones,   (*Cratægus*) HAWTHORN.

    Fruit with 3 to 5 parchment-like pips.

      Leaves pinnate: fruit berry-like, scarlet when ripe.   (*Pyrus*, § *Sorbus*) MOUNTAIN-ASH.

      Leaves simple.

        Flowers small in compound cymes: fruit small, berry-like, black or

          dark red, mawkish,   (*Pyrus*, § *Adenòrachis*) CHOKEBERRY.

        Flowers large in simple clusters or umbels: fruit fleshy.

          Petals tinged with red or rose: fruit sunk in at both ends, (*Pyrus*, § *Malus*) APPLE.

          Petals white: fruit tapering into the stalk,   (True *Pyrus*) * PEAR.

Cells of the fruit parchment-like and many-seeded,   (*Cydònia*) * QUINCE.

Cherry. *Prunus*, § *Cèrasus*, &c.

* Flowers, like those of Plums, two or more together on separate footstalks from separate lateral buds, appearing at the same time with the leaves.

1. CULTIVATED CHERRY: several varieties are commonly cultivated of the European,   *P. Cèrasus*.

2. WILD RED CHERRY. A small tree, with bright-green narrow leaves, and small light-red sour fruit. Common in rocky woods, &c.   *P. Pennsylvánica*.

   * * Flowers in hanging racemes, appearing after the leaves, late in spring. Wild species.

n. CHOKE CHERRY. Shrub or small tree, with gray branches, broad and sharply serrate leaves, and astringent dark crimson fruit, ripe in summer.   *Virginiàna*.

4. WILD BLACK CHERRY. Shrub or large tree, with reddish-brown bark on the branches, oblong or lance-oblong leaves with short and blunt teeth, and purplish-black vinous fruit, ripe in autumn.
P. serótina.

Plum. *Prunus.*

All are cultivated, except the Beach Plum ; but No. 2 is also wild; so is No. 3 in the Southwest.

1. COMMON PLUM (*P. doméstica*), with all its varieties, probably came from the BULLACE PLUM (*P. insitítia*), and that perhaps from the thorny SLOE (*P. spinòsa*).

2. WILD (RED and YELLOW) PLUM : well known for its very juicy pulp in a (red or partly yellow) tough skin; leaves coarsely serrate.
P. Americàna.

3. CHICKASAW PLUM : with lance-shaped finely serrate leaves, and small red, thin-skinned, cherry-like fruit. S.
P. Chicasa.

4. BEACH PLUM. A low bush on the sea-coast, with the leaves downy beneath, and a small purple or crimson fruit.
P. marítima.

Meadow-Sweet. *Spiræa.*

Calyx 5-cleft. Petals 5, broad or roundish. Pistils commonly 5, making little pods (follicles) with 2 or few seeds in each. Nos. 1, 2, 4, and 6 are wild species, but also cult. in gardens and grounds.

* Shrubs, with white flowers, except No. 2.

1. COMMON MEADOW-SWEET. Smooth, 2° or 3° high ; leaves oblong or lance-oblong and wedge-shaped; flowers in a crowded panicle, sometimes pale flesh-color. Wet grounds.
S. salicifòlia.

2. DOWNY M., or HARDHACK. Leaves coated with wool beneath; flowers rose-color.
S. tomentòsa.

3. ITALIAN M., or MAYWREATH. Smooth ; stems 3° or 4° long, recurved; leaves small, spatulate, entire; flowers small, in umbels on short leafy shoots. Cult.; fl. in spring.
S. hypericifòlia.

4. NINEBARK M. Smoothish, 4° to 10° high; branches recurving; leaves rounded, 3-lobed ; flowers in umbels, in spring; pods 3 to 5, bladdery, turning purplish. Old bark of stems peeling off in thin layers. Rocky banks, N. & W., and cultivated.
S. opulifòlia.

5. SORB-LEAVED M. Smooth, 3° to 6° high; leaves pinnate; leaflets oblong-lance-shaped, pointed, cut-toothed; flowers in a large panicle, in spring. Cultivated.
S. sorbifòlia.

* * Herbs, with perennial roots, and interruptedly pinnate leaves, and flowers in a crowded compound cyme, on a long naked stalk. All but No. 6 are foreign species.

6. QUEEN-OF-THE-PRAIRIE M. Smooth; leaflets 3 to 7 and some little ones; end-leaflet very large, parted and cleft; flowers peach-blossom-color, in summer. W. and cult.
S. lobàta.

7. ENGLISH M. Leaves smaller than in the last, white-downy beneath; flowers white.
S. Ulmària.

8. DROPWORT M. Smooth; leaflets 9 to 21, besides the minute ones, linear-oblong, much cut ; cymes of a few slender branches; flowers white, single or double.
S. filipéndula.

Indian-Physic. *Gillènia.*

Calyx narrow or club-shaped, 5-toothed. Petals 5, lance-shaped, rather unequal, white or pale rose. Stamens 10 to 20, short. Pistils and little pods 5. — Herbs, with perennial roots, and leaves of three cut-toothed thin leaflets. Flowers in a loose corymb or panicle, in summer.

1. COMMON INDIAN-PHYSIC (or BOWMAN'S ROOT). Leaflets oblong; stipules small and entire. W. and cultivated in gardens.
G. trifoliàta.

2. WESTERN I. (or AMERICAN IPECAC). Leaflets lance-shaped, more cut than in the last, as are the large stipules. W.
G. stipulàcea.

Avens. *Geum.*

Calyx bell-shaped or flattish, 5-cleft, and with 5 additional little lobes between. Petals 5. Stamens many. Pistils many in a head, making akenes, which are tipped with the style, remaining as a long, naked or hairy tail. Perennial herbs: flowers single or somewhat corymbed. — In all our common species the style is jointed and hooked round in the middle.

* Upper and mostly hairy joint of the style falling off, leaving the lower and smooth portion, which remains hooked at the end: flowers rather small: root-leaves mostly interruptedly pinnate; stem-leaves or lobes 3 to 5. Dry woods and fields.

1. WHITE AVENS. Smoothish or downy; petals white, as long as the calyx, akenes bristly. *G. album.*
2. VIRGINIAN A. Bristly-hairy, stouter than the last; petals greenish-white, shorter than the calyx; akenes smooth. *G. Virginiànum.*
3. YELLOW A. Rather hairy, large; petals yellow, longer than the calyx. *G. strictum.*

* * Upper joint of the style persistent and feathered with long hairs; flowers rather large, nodding.

4. WATER A. Root-leaves with a large and rounded-lobed end-leaflet, and some very small ones below; stem-leaves few, 3-cleft or of 3 small leaflets; petals not spreading, somewhat notched at the broad summit, purplish. — Wet banks of streams. *G. rivàle.*

Cinquefoil. *Potentilla.*

Calyx open or flat, 5-parted, and with 5 additional outside lobes alternate with the others, making 10. Petals 5. Stamens many. Pistils many in a head, on a dry receptacle, making seed-like akenes, the styles falling off.

* Leaves palmate. Herbs, with yellow flowers.

1. NORWAY CINQUEFOIL. Erect, coarse, hairy; leaflets 3, obovate, cut-toothed. Fields. *P. Norvègica.*
2. CANADA C. Runner-like stems decumbent or spreading; leaflets 5, obovate-oblong; peduncles long, axillary, 1-flowered. Fields and banks. *P. Canadénsis.*
3. SILVERY C. Low, with spreading branches, white-woolly, as are the 5 leaflets beneath. *P. argéntea.*

* * Leaves pinnate. Herbs (except No. 5): receptacle of the fruit hairy.

4. SILVER-WEED. Creeping, sending up leaves of 9 to 19 cut-toothed leaflets, besides little ones interposed, silvery-white beneath, and single long-stalked yellow flowers. Wet banks, N. *P. Anserìna.*
5. SHRUBBY C. Shrub very bushy, 2° to 4° high; leaflets 5 or 7, crowded near the end of the short footstalk, lance-oblong, entire, silky beneath; flowers yellow. Bogs. *P. fruticòsa.*
6. MARSH C. Stems ascending from a scaly creeping base; leaflets 5 or 7, crowded, serrate, lance-oblong; flowers dull purple. Cold bogs, N. *P. palùstris.*

Bramble. *Rubus.*

Calyx open, deeply 5-cleft. Petals 5. Pistils many; their ovaries ripening into little berry-like grains (or rather *drupelets*), making a kind of compound berry. — Rather shrubby or herbaceous perennials.

§ 1. RASPBERRY. Fruit falling from the dry receptacle, usually with the grains lightly cohering.

* Leaves simple, lobed: flowers large and showy: petals spreading.

1. PURPLE FLOWERING-RASPBERRY. Bristly and clammy with odorous brownish glands; leaves rounded, with 3 or 5 pointed lobes; flowers in a corymb, rose-purple; fruit flat. Rocky banks, N. Fl. summer. *R. odoràtus.*
2. WHITE FLOWERING-R. Like No. 1, but the flowers white and smaller. N. W. & cult. *R. Nutkànus.*

* * Leaflets 3 or 5, white-downy beneath: flowers small: petals white, erect.

8. GARDEN RASPBERRY. Stems with some slender hooked prickles as well as bristles; petals shorter than the calyx; fruit red, &c., the grains minutely downy. Cult. *R. Idæus.*

4. WILD RED R. Stems very bristly; petals as long as the calyx; fruit pale red, very tender. Very common N. *R. strigòsus.*

5. BLACK R. (or THIMBLEBERRY). Plant glaucous all over; the long recurved stems and stalks beset with hooked prickles; fruit dark purple. Borders of woods and fields. *R. occidentàlis.*

§ 2. BLACKBERRY. Fruit of large grains, remaining on the juicy receptacle, black or dark purple when ripe: petals white, spreading; leaflets 3 or 5.

6. HIGH BLACKBERRY or BRAMBLE. Stems mostly erect, angular, bearing stout curved prickles; young shoots hairy and glandular; leaflets ovate or oblong, pointed, downy underneath and prickly on the midrib; flowers large, in racemes; fruit large, sweet. *R. villòsus.*

7. LOW B. (or DEWBERRY). Stems long, trailing; leaves smaller and nearly smooth; flowers fewer, and the large sweet fruit ripe earlier than in the last. Sterile or rocky ground. *R. Canadénsis.*

8. SAND B. Stems low, but erect, with stout hooked prickles; leaflets wedge-obovate, whitish-woolly beneath; fruit sweet. Sandy soil. New Jersey & S. *R. cuneifòlius.*

9. RUNNING SWAMP-B. Stems slender, creeping, hooked-prickly; leaves nearly evergreen, shining, obovate; flowers small; fruit of few grains, reddish until ripe, sour. Wet woods, N. *R. hispidus.*

Rose. *Rosa.*

Calyx with an urn-shaped hollow tube (Fig. 360), bearing 5 leafy lobes at the top, 5 petals and many stamens, and within enclosing many pistils attached to its walls. The ovaries ripen into bony and hairy akenes, and the calyx makes a fleshy or pulpy, red and berry-like fruit (*hip*). — Shrubs, with pinnate leaves of 3 to 9 leaflets. (Stigmas just rising to the mouth of the calyx, except in No. 1.)

* Wild Roses. But No. 1 is cultivated, especially in double-flowered varieties, and the Sweet-Brier, which came from Europe, is also kept in gardens, for its sweet-scented leaves. Flowers in all bright rose-color.

1. PRAIRIE ROSE. Stems climbing high, prickly; leaflets 3 or 5, large; petals deep rose-color turning pale; styles cohering together, and projecting out of the tube of the calyx; flowers in corymbs, scentless, in summer. Edges of prairies and thickets; W. and cult. *R. setigera.*

2. SWEET-BRIER R. (or EGLANTINE). Stems climbing, and with stout hooked prickles; leaflets 5 or 7, roundish, downy and bearing russet fragrant glands beneath; hip pear-shaped. Road-sides, gardens, &c. *R. rubiginòsa.*

3. SWAMP R. Stems erect, 4° to 7° high, with hooked prickles; leaflets dull, 5 to 9; flowers in corymbs; hips rather bristly, broader than long. *R. Carolina.*

4. LOW WILD R. Stems 1° to 3° high, with mostly straight prickles; leaves smooth and commonly shining; flowers single or 2 to 3 together; hips as in the last Common. *R. lùcida.*

5. BLAND R. Low, pale or glaucous, with few or no prickles; calyx and globular hips very smooth. Rocks: flowering early in summer. N. *R. blanda.*

* * Cultivated species are very numerous and much mixed. The commonest are: —

CINNAMON ROSE, *R. cinnamòmea.* DAMASK R., *R. Damascèna.*
SCOTCH or BURNET R., *R. spinosissima.* CABBAGE or HUNDRED-LEAVED R., *R. centifòlia.*

Moss R., *R. centifòlia*, var. *muscòsa*.
White R., *R. alba*.
Yellow R., *R. lùtea*.

China R., *R. Indica*.
Cherokee R. at the South, *R. lœvigàta.*
Multiflora R., *R. multiflòra.*

Hawthorn. *Cratœgus.*

Calyx with a globular or pear-shaped tube coherent with the 2- to 5-celled ovary, making a pome with as many one-seeded *stones.* Petals 5, roundish. Styles 2 to 6. Thorny small trees or shrubs. Flowers in spring, mostly in corymbs, white, or with a red variety of the cultivated.

1. English Hawthorn (or White Thorn). Leaves obovate, with a wedge-shaped base, lobed and cut; styles 2 or 3; fruit small, coral-red. Cult. for hedges and ornament. *C. Oxyacántha.*
2. Washington H. Leaves broadly ovate, truncate or a little heart-shaped at the base, often cleft or cut; styles 5; fruits coral-red, not larger than peas. S. *C. cordàta.*
3. Scarlet-fruited H. Smooth; leaves round-ovate, thin, toothed or cut, on slender stalks; fruit scarlet, oval, ½' in diameter. *C. coccinea.*
4. Pear H. (or Blackthorn). Downy, at least when young; leaves thickish, oval, ovate, or wedge-obovate, narrowed into a short or margined footstalk; flowers large; fruit large, crimson, or orange-red, eatable. *C. tomentòsa.*
5. Cockspur H. Smooth; leaves wedge-obovate or inversely lance-shaped, merely toothed above the middle, thick, shining; fruit dark red; thorns very long. *C. Crus-gálli.*
6. Summer H. Rather downy; leaves obovate or wedge-shaped, often cut; flowers few (2 to 6); fruit rather pear-shaped, yellowish or reddish. S. *C. flava.*

Apple. *Pyrus,* § *Malus.*

1. Common Apple. Leaves ovate, serrate, downy beneath; flowers white tinged with pink. Everywhere cultivated. *P. Malus.*
2. Siberian Crab-A. Leaves ovate, serrate, smooth; calyx smooth. Cult. occasionally. *P. baccàta.*
3. American Crab-A. Leaves broadly ovate or heart-shaped, cut-toothed or somewhat lobed, smoothish; flowers rose-color, sweet-scented; fruit greenish, fragrant (Fig. 361). Common. W. *P. coronària.*

Mountain-Ash or Rowan Tree. *Pyrus,* § *Sorbus.*

Both the wild and the foreign species are planted for the beauty of their bright scarlet fruits, in broad compound cymes, ripe in autumn. Fl. white, summer.

1. American M. Leaflets 13 to 15, lance-shaped, taper-pointed, smooth. Wild, N. *P. Americàna.*
2. European M. Leaflets shorter, broader, paler, and not pointed; fruit larger. *P. aucupària.*

Quince. *Cydònia.*

1. Common Quince. Flowers single at the tips of the branches, white; lobes of the calyx leaf-like and downy, as well as the ovate entire leaves; fruit pear-shaped. Cult. *C. vulgàris.*
2. Japan Quince. Shrub, hardly of the same genus, for the flowers are on side spurs of the thorny branches, earlier than the smooth leaves; calyx top-shaped, with short lobes; petals large and red; fruit like a small apple, very hard. Cultivated for ornament. *C. Japónica.*

34. CAROLINA-ALLSPICE FAMILY. Order CALYCANTHACEÆ.

A small family of a few rather curious shrubs, with opposite leaves; represented by the

Carolina-Allspice. *Calycánthus.*

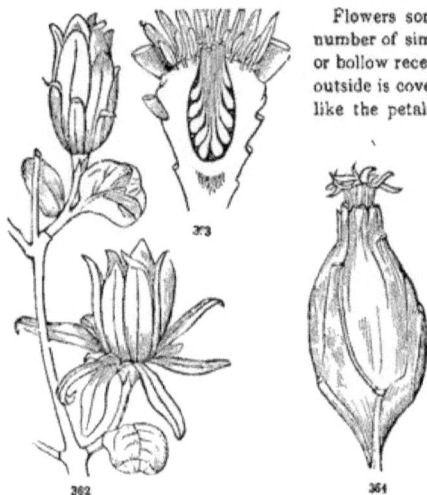

Flowers somewhat on the plan of the rose, having a large number of simple pistils contained in a sort of closed calyx-cup, or hollow receptacle, and attached to its inner surface. But the outside is covered with sepals or calyx-lobes, which are colored like the petals (brown-purple); these are many and narrow, in several rows. Stamens many, on the top of the cup; filaments hardly any; anthers long, tipped with a point. Ovaries making large akenes, enclosed in the large and dry hip. Seed-leaves of the embryo rolled up. Shrubs, with rather aromatic bark, &c., and opposite entire leaves, without any stipules. Flowers large, when bruised giving out a fragrance resembling that of strawberries. Wild in the Southern States, especially in and near the mountains; and also cultivated, especially the first species.

362. Flowering branch of Carolina Allspice. 363 Half of a calyx-cup of the same, cut through lengthwise. (Compare it with a Rose, Fig. 360.) 364. A ripe fruit or hip.

1. COMMON C.   Leaves oval or roundish, downy beneath.   Commonly cult. in gardens.   *C. flóridus.*
2. SMOOTH C.   Leaves oblong, smooth, green both sides; flowers smaller.   *C. lævigàtus.*
3. GLAUCOUS C.   Leaves oblong- or lance-ovate, pointed, glaucous or whitened beneath.   *C. glaucus.*

35. LYTHRUM FAMILY. Order LYTHRACEÆ.

Herbs with entire and mostly opposite leaves, and no stipules; the calyx tubular or cup-shaped, bearing from 4 to 7 petals and 4 to 14 stamens on its throat, and enclosing the many-seeded ovary and thin pod. Between the 4 to 7 teeth of the calyx are as many additional projections or supernumerary teeth. Style one.

Flowers regular, or nearly so.
    Calyx cylindrical, several-ribbed or angled: petals 4 to 7, rather unequal: stamens
       twice as many as the petals: pod 2-celled,   (*Lythrum*) LYTHRUM.*
    Calyx short bell-shaped: petals 5: stamens 10 or 14, long and protruded: pod with
       3 to 5 cells: leaves often whorled,   (*Nesæa*) NES.ÆA.
Flowers with an irregular tubular calyx, spurred or projecting at the base on the upper
    side. Very unequal petals, and 12 unequal stamens in two sets. Pod few-seeded,
       bursting through one side of the calyx,   (*Cúphea*) CUPHEA.

' Sometimes called *Loosestrife*; but this name properly belongs to plants of another family.

36. EVENING-PRIMROSE FAMILY. Order ONAGRACEÆ.

Herbs, or sometimes shrubs, known by having the parts of the blossom in fours, the tube of the calyx coherent with the 4-celled ovary, and often prolonged beyond, its summit bearing 4 petals, and 4 or 8 stamens. Style 1, slender: stigmas generally 4. In green-house cultivation we have several species of FUCHSIA, well known for their pretty hanging flowers, the smaller kinds called *Ladies' Eardrop.* The showy part is a colored (generally red) calyx, its 4 lobes longer than the purple petals. Fuchsias are shrubs; the rest of the family are herbs. CLARKIA, known by the long-clawed petals, and broad petal-like stigmas, is sometimes cultivated, and so are several Evening-Primroses. The commonest wild plants of the family are EVENING-PRIMROSES and WILLOW-HERBS.

Evening-Primrose. *Œnothèra.*

Calyx with the tube continued on beyond the ovary, bearing 4 narrow lobes turned down, 4 generally obcordate petals, and 8 stamens. — Several species are cultivated more or less commonly in flower-gardens. The following are common wild, and have yellow flowers, in summer.

1. COMMON E. Tall; leaves lance-shaped; flowers in a spike, opening at sunset or in cloudy weather, sweet-scented; pod cylindrical; root biennial. Fields, &c. *Œ. biénnis.*

2. LOW E. Stems several from a perennial root, 1° to 3° high; flowers large, opening in sunshine; pods rather club-shaped, and 4-winged, stalked. W. & S. *Œ. fruticòsa.*

3. SMALL E. Stems ½° to 1° high; flowers small, ½' wide, open in sunshine; pods club-shaped, scarcely stalked, strongly 4-angled. Fields, &c. *Œ. pùmila.*

Willow-herb. *Epilòbium.*

Calyx with its tube not continued beyond the ovary. Petals 4, purple or whitish. Stamens 8. Pod long and slender, many-seeded; the seeds bearing a long tuft of downy hairs.

1. GREAT W. Stem simple, 4° to 7° high; leaves lance-shaped; flowers showy, pink-purple, in a long loose spike; petals on claws, widely spreading; stamens and style turned down. Rich ground, especially where it has been burned over or newly cleared. *E. angustifòlium.*

2. SMALL W. Branching, 1° to 2° high; leaves lance-oblong, commonly purple-veined; flowers very small; petals purplish. Wet places, everywhere. *E. coloràtum.*

37. CACTUS FAMILY. Order CACTACEÆ.

Fleshy and generally prickly plants, without any leaves, except little scales or points, of very various and strange shapes, generally the petals and always the stamens very numerous, and on the one-celled ovary, which in fruit makes a berry. Being house-plants (with one exception) they must here be passed by, merely mentioning the

PRICKLY-PEAR CACTUS, which grows in dry sandy or rocky places, southward, and consists of flat and rather leaf-like rounded joints of stem, growing one out of another, prickly at the buds, and bearing yellow flowers of rather few petals; the ovary making a large berry full of sweet and eatable pulp. *Opùntia vulgàris.*

38. GOURD FAMILY. Order CUCURBITACEÆ.

Succulent or tender herbs, with alternate and radiate-veined leaves, and with tendrils. Flowers commonly monœcious, in the axils. Fertile flowers with the tube of the calyx coherent with the ovary. Petals often united with each other into a monopetalous corolla, and united with or borne on the cup of the calyx. Stamens generally 3, and more or less connected by their anthers or their filaments, or by both ; the anthers curiously contorted. Fruit a pepo (224), berry, or pod. Seeds large and flat ; the whole kernel is an embryo. The most important plants of the family are those cultivated.

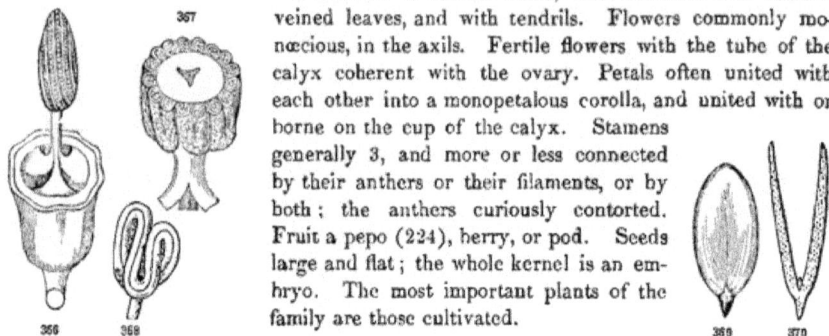

366. Staminate flower of a Squash, with the corolla and upper part of the calyx cut away, to show the united stamens.   367. The latter, enlarged, and the mass of anthers cut across   368. Separate stamen of a Melon, enlarged, showing the long and contorted anther.   369. Embryo of Squash.   370. Section of same, a little enlarged, seen edgewise.

Petals united into a large, bell-shaped, 5-lobed, yellow corolla. Stamens with three filaments united into a tube, except at the bottom: the anthers also firmly grown together; the turns of their long cells parallel, running straight up and down. Style 1: stigmas 3, each 2-lobed. Fruit large, firm-fleshy. Seeds with a blunt edge,   (*Cucúrbita*) *GOURD, i. o.

Petals united only at the base or separate. Anthers loosely crooked.   [SQUASH and PUMPKIN.
    Ovary and fruit many-seeded. Anthers and filaments 3, separate or separable.
      Petals white, with greenish veins. Peduncles very long. Fruit with a hard or
        woody rind variously shaped,   (*Lagenària*) *BOTTLE-GOURD.
      Petals yellow. Calyx with a bell-shaped cup. Seeds pointed and sharp-edged.
        Fruit narrow, rough-pimpled when young,   (*Cùcumis satìvus*) *CUCUMBER.
        Fruit thick, smooth, sweet. Fertile flowers perfect,   (*Cucumis Melo*) *MUSKMELON.
      Petals buff or cream-color. Calyx with hardly any cup. Leaves much cut.
        Fruit large and smooth, sweet. Seeds thick-edged, smooth, (*Citrùllus*) *WATERMELON.
        Fruit a rough, reddish berry. Seeds wrinkled,   (*Momórdica*) *BALSAM-APPLE.
    Ovary and fruit one-seeded or 4-seeded. Small-flowered climbers, wild in this country.
      Corolla of the sterile flowers 6-parted, white. The long racemes rather pretty in
        cultivation. Fruit an oval, weak-prickly, bladder-like pod, bursting
        at the top, and containing 2 fibrous-netted cells, with 2 large seeds in
        each. Leaves sharply 5-lobed,   (*Echinocýstis*) BLADDER-CUCUMBER.
      Corolla of the wheel-shaped sterile flowers 5-lobed, greenish-white. Fruit a
        small, ovate, 1-seeded, prickly-barbed bur. Leaves 5-angled, (*Sícyos*) BUR-CUCUMBER.

39. PASSION-FLOWER FAMILY. Order PASSIFLORACEÆ.

This small family of tendril-bearing vines, with alternate palmately-lobed leaves, is mainly represented by the

Passion-Flower. *Passiflòra.*

Sepals 5, united at the base. Petals 5, accompanied by a *crown* or ring formed of a double or triple fringe, inserted on the base of the calyx. Stamens 5, mona-delphous; the filaments making a long sheath to the slender stalk of the ovary: this is one-celled and becomes an eatable berry, with many seeds in 3 or 4 rows on its walls. The species are mostly South American; and some large-flowered and handsome ones are cultivated in hot-houses. The early missionaries fancied that they found in these flowers emblems of the implements of our Saviour's passion; the fringe representing the crown of thorns; the large anthers fixed by their middle, hammers; and the 5 styles (tapering below and with large-headed stigmas), the nails. We have two wild species, common S. and W.

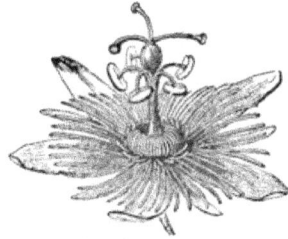

371. Passion-Flower No. 1, enlarged.

1. SMALL P. Leaves bluntly 3-lobed, otherwise entire; flowers greenish-yellow, 1' wide. *P. lùtea.*
2. MAYPOP P. Leaves 3-cleft, the lobes serrate; flowers 2' broad, white, with a triple flesh-colored and purple crown; fruit like a hen's egg in shape and size. *P. incarnàta.*

40. CURRANT FAMILY. Order GROSSULACEÆ.

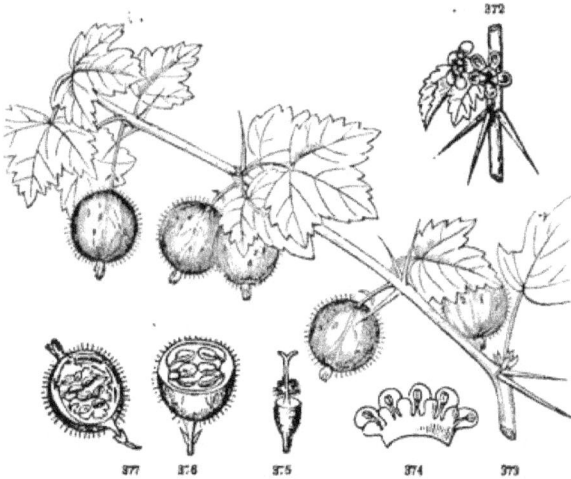

Consists of the Currants and Gooseberries, which belong to the same botanical genus. Shrubs, with alternate rounded and radiate-veined leaves; the tube of the calyx coherent with the one-celled ovary, and continued above it into a cup which is often colored, like a corolla, and bears the 5 little petals and 5 stamens. Seeds many, with a pulpy outer coat, borne upon the walls of the berry on two thickened lines (parietal placentas).

Garden Gooseberry: 372. with flowers; 373. with fruit. 374. Cup of the calyx laid open, bearing the 5 little petals and stamens. 375. The pistil. 376. Young berry cut across. 377. Young berry divided lengthwise.

Gooseberry. *Ribes,* § *Grossulària.*

Stems generally armed with thorns under the clusters of leaves, and sometimes with scattered prickles. Peduncles bearing single or few flowers.

11

1. GARDEN GOOSEBERRY. Thorns large; flower-stalks short; berry bristly or smooth. *R. Uva-crispa.*
2. PRICKLY WILD G. Thorns slender or none; flowers greenish, long-stalked; stamens and style not projecting; berry prickly; leaves downy. Woods, N.   *R. Cynósbati.*
3. SMALL WILD G. Thorns very short or none; flowers purplish or greenish, very short-stalked; stamens and 2-cleft style a little projecting; berry small, smooth. Low grounds, N.   *R. hirtéllum.*
4. SMOOTH WILD G. Thorns stout or none; flowers greenish, on slender stalks; stamens and the two styles very long and projecting (½′ long); berry smooth. Woods, common W.   *R. rotundifólium.*

Currant. *Ribes.*

Stems neither thorny nor prickly. Flowers in racemes, appearing in early spring. Berries small.

1. RED CURRANT. Leaves rounded heart-shaped and somewhat lobed; racemes from lateral separate buds, hanging; flowers flat, greenish or purplish; berry smooth, red, and a white variety. Gardens, &c. Wild on Mountains, N.   *R. rúbrum.*
2. FETID C. Stems reclined; leaves deeply heart-shaped, 5-lobed; racemes erect; flowers greenish, flattish; pale red berry and its stalk bristly, strong-smelling. Cold woods, N.   *R. prostrátum.*
3. WILD BLACK C. Leaves on long foot-stalks, slightly heart-shaped, sharply lobed, sprinkled with dots both sides; racemes rather drooping; flowers oblong, yellowish-white; berries oblong, black, rather spicy. Wooded banks.   *R. flóridum.*
4. GARDEN BLACK C. Leaves on shorter footstalks, less dotted; racemes looser, and black berries larger than in No. 3. Gardens.   *R. nigrum.*
5. MISSOURI or BUFFALO C. Leaves smooth; racemes with leafy bracts; flowers (calyx) long and tubular, bright yellow, spicy-fragrant. Cultivated for ornament.   *R. aúreum.*

41. STONECROP FAMILY. Order CRASSULACEÆ.

Herbs with thick and fleshy leaves (except in one peculiar plant of the family, viz. the Ditchwort); the flowers remarkable for being perfectly regular and symmetrical throughout, i. e. having the sepals, petals, and pistils all of the same number and all separate, or nearly so (except in Ditchwort); the stamens also of the same number, or just twice as many. Pods containing few or many seeds. Mostly small plants: several are found in gardens.

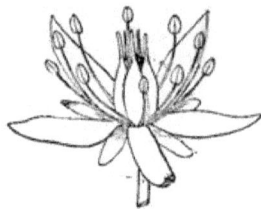
378. Flower of Stonecrop.

Flowers with petals, and their pistils entirely separate from each other.
 Sepals, narrow petals, and pistils 4 or 5. Stamens 8 or 10,   (*Sedum*) STONECROP.
 Sepals, petals, and pistils 6 to 20. Stamens 12 to 40,   (*Sempervivum*) HOUSELEEK.
Flowers with 5 sepals, no petals, and 5 pistils grown together below. Leaves thin, lance-
 shaped,   ▪   (*Pénthorum*) DITCHWORT.

Stonecrop or Orpine. *Sedum.*

1. MOSSY STONECROP. Small and creeping, moss-like; the stems thickly covered with little ovate thick and closely sessile leaves; flowers yellow. Cultivated for garden edging, &c.   *S. acre.*

2. THREE-LEAVED S. Stems spreading, 3' to 5' high; leaves wedge-obovate or oblong, the lower ones in whorls of 3; the earliest flower with the parts in fives, the rest generally in fours; petals white. Rocky woods, S. and W. and in gardens. *S. ternátum.*

3. HANDSOME S. Stems 4' to 12' high; leaves thread-shaped; flowers crowded; petals rose-purple. Rocky places, S. W. and cultivated. *S. pulchéllum.*

4. GREAT S. or LIVE-FOR-EVER. Stems 2° high; leaves oval; flowers in a close compound cyme, purple. Gardens. *S. Telèphium.*

42. SAXIFRAGE FAMILY. Order SAXIFRAGACEÆ.

Herbs, or in the case of Hydrangea, &c. shrubs, differing from the last in having the pistils fewer than the petals, and generally more or less united with each other and with the tube of the calyx. Petals 5 (rarely 4), on the calyx. Stamens 5 or 10, or in Mock-Orange many.

Herbs. Leaves generally alternate. Petals 5. Styles only 2.
  Stamens 10, short. Petals entire. Calyx deeply 5-cleft. Pod 2-beaked or pods 2, many-seeded,   *(Saxifraga)* SAXIFRAGE.
  Stamens 5. Petals small, entire (greenish or purplish), between the short lobes of the bell-shaped calyx. Pod 1-celled, 2-beaked, many-seeded. Flowers in a long panicle,   *(Heúchera)* ALUM-ROOT.
  Stamens 10, short. Petals pinnatifid, whitish, slender. Styles and pod short, one-celled, the latter few-seeded at the bottom, opening across the top. Stem 2-leaved below the slender raceme,   *(Mitélla)* MITREWORT.
  Stamens 10, and the 2 styles much longer than the slender-clawed petals. Pod slender, few-seeded at the bottom. Flowers white in a short raceme on a naked scape,   *(Tiarélla)* FALSE-MITREWORT.
Shrubs. Leaves opposite. Tube of the calyx coherent with the ovary. Seeds many.
  Flowers small, in compound cymes; some of the marginal ones generally large and neutral (Fig. 169), or in cultivation nearly all the flowers becoming so. Petals 4 or 5. Stamens 8 or 10. Styles 2, diverging, and between them the little pod opens,   *(Hydrángea)* HYDRANGEA.
  Flowers large, somewhat panicled. Petals 4 or 5, white, showy. Stamens 20 or more. Styles 3 to 5, united below: pod with as many cells, very many-seeded,   *(Philadélphus)* MOCK-ORANGE.

Saxifrage. *Saxifraga.*

1. EARLY SAXIFRAGE. Leaves all clustered at the root, obovate, toothed; scape 4' to 9' high, many-flowered; flowers white, in early spring. Damp rocks. *S. Virginiénsis.*

2. SWAMP S. Leaves all at the root, lance-oblong, 3' to 8' long; scape 1° or 2° high, clammy, bearing many small clustered greenish flowers. Bogs and wet ground, N. *S. Pennsylvánica.*

· Hydrangea. *Hydrángea.*

1. GARDEN HYDRANGEA. Leaves very smooth; flowers mostly large neutral ones, blue, purple, or pink. A well-known garden and house plant. *H. Horténsia.*

  WILD H. Leaves thin, nearly smooth, sometimes heart-shaped; flowers mostly perfect, white.
   H. arboréscens.

Mock-Orange (or **Syringa**). *Philadélphus.*

1. Common M. or Syringa. Flowers cream-colored, fragrant, in large panicles; styles separate. Cultivated.   *P. coronàrius.*

2. Scentless M. Flowers larger and later than in the first, few on the spreading branchlets, pure white. Cultivated; also wild S. Leaves tasting like cucumbers.   *P. inodòrus.*

43. PARSLEY FAMILY. Order UMBELLIFERÆ.

Herbs with small flowers in compound umbels, the 5 petals and 5 stamens on the top of the ovary, with which the calyx is so incorporated that it is not apparent, except sometimes by 5 minute teeth. Styles 2. Fruit dry, 2-seeded, splitting when ripe into two akenes. Stems hollow. Leaves generally compound, decompound, or much cut. Some species are aromatic, having a volatile oil in the seeds: most, but not all, of these are harmless. Others contain a deadly poison in the roots and leaves. The deadly poisonous sorts are marked †: the most deadly is the *Water-Hemlock*, also called *Musquash-root*, and *Beaver-Poison.* — The kinds in this large family are known by their fruit, and are too difficult for the beginner. The principal common kinds are merely enumerated in the following key. (Fig. 148 shows the compound umbel in Caraway, a good and familiar example of the family.)

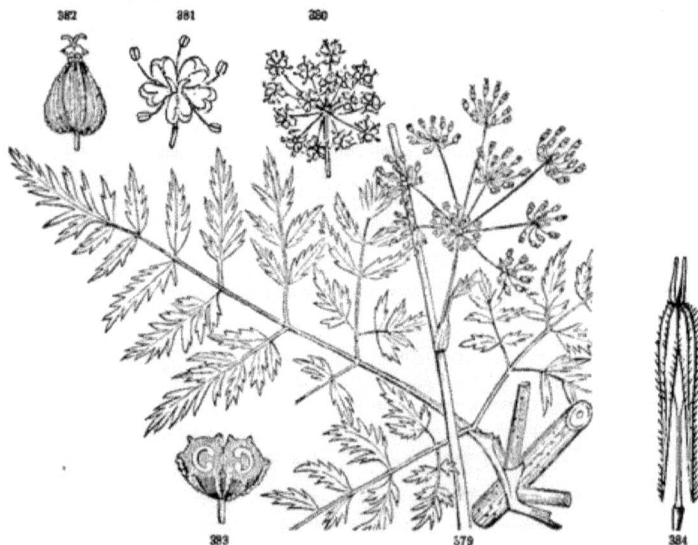

379. Part of Stem, leaf, umbel, &c of Poison-Hemlock. 380 A separate umbellet. 381. A flower magnified. 382 A fruit 363. Lower half of it cut off. 384. Fruit of Sweet Cicely ; the two long akenes separating.

Seeds flat on the inner face, where the two akenes or parts of the fruit join.

Fruit covered all over with hooked prickles,   (*Sanicula*) SANICLE.

Fruit prickly on the ribs only. Umbel becoming concave,   (*Daucus*) *CARROT.

Fruit not prickly, but winged on the margin.

    Flowers yellow, all alike,   (*Pastinàca*) *PARSNIP.

        Flowers white, the outer corollas larger,   (*Heraclèum*) COW-PARSNIP.

        Flowers white or whitish, all alike.

            Akenes 5-ribbed on the back. Leaves simply pinnate,   (*Archèmora*) COWBANE.†

            Akenes 3-ribbed on the back. Leaves decompound,   (*Angèlica*) *ANGELICA.

Fruit not prickly, winged on all sides,   (*Levisticum*) *LOVAGE.

Fruit neither prickly nor winged.

    Flowers yellow. Plant sweet-aromatic; leaflets long and slender,   (*Fœniculum*) *FENNEL.

    Flowers white.

        Umbels with neither involucre nor involucels.

            Divisions of the leaves very slender,   (*Carum*) *CARAWAY.

            Divisions or leaflets wedge-shaped,   (*Apium*) *CELERY.

        Umbels with 3-leaved involucels, but no involucre,   (*Æthùsa*) FOOL'S-PARSLEY.

        Umbels with both involucres and involucels.

            Leaves decompound, finely divided,   (*Petroselìnum*) *PARSLEY.

            Leaves 2 or 3 times compound; leaflets coarse,   (*Cicùta*) WATER-HEMLOCK.†

            Leaves simply pinnate,   (*Sium*) WATER-PARSNIP.†

Seed grooved or hollowed down the whole length of the inner face. (Flowers white.)

Herbage rather unpleasant-scented: leaves decompound, finely cut, (*Conium*) POISON-HEMLOCK.†

Herbage, fruit, &c. sweet-scented.

    Fruit narrow-oblong, ribbed,   (*Chœrophỳllum*) CHERVIL.

    Fruit long, tapering downwards,   (*Osmorrhìza*) SWEET-CICELY.

Seed and fruit curved in at the top and bottom, or kidney-shaped, strong-scented.

    Flowers white,   (*Coriándrum*) *CORIANDER.

44. ARALIA FAMILY. Order ARALIACEÆ.

Much like the last, but often shrubs or trees; the styles almost always more than two, and the fruit becoming berry-like. Also the umbels are not regularly compound, but either simple or panicled. Flowers often polygamous. Here belongs the true or English Ivy, with evergreen simple leaves, which thrives in some places in northern exposures; also the following wild plants.

Aralia. *Aràlia.*

Petals, stamens, and styles 5. Flowers white or greenish in summer. Berries black. Herbage, roots, &c. aromatic. Leaves compound or decompound, large.

1. PRICKLY A. or ANGELICA-TREE. Shrub or low tree with a stout simple stem, very prickly; leaves very large; leaflets ovate; umbels many in a large panicle. S. and cult.   *A. spinòsa.*

2. BRISTLY A. Stem 1° high, bristly below, woody at the base; leaves twice pinnate; umbels few, corymbed. Rocky woods. N.   *A. hispida.*

3. SPIKENARD A. A stout spreading herb; with thick sweet-spicy roots; leaves very large and decompound; leaflets somewhat heart-shaped; umbels many, panicled. Rich woods. *A. racemòsa.*

4. Sarsaparilla A.   Roots very long and slender, horizontal (used as a substitute for sarsaparilla); the compound long-stalked leaf, and the naked flower-stalk bearing few umbels, rising separately from the ground.   Moist woods.   *A. nudicaùlis.*

Ginseng. *Aralia, § Ginseng.*

Styles 2 or 3.   Flowers white.   Berries red or reddish when ripe.   Low berbs with simple stems, bearing at the top a whorl of leaves and one long-stalked umbel.

1. True Ginseng.   Root long and large, warm-aromatic; leaflets 5.   Rich woods, N.   *A. quinquefòlia.*
2. Dwarf G. (or Groundnut).   Root round, sharp-tasted; leaflets 3 or 5; stem 4' to 6' high.   Damp woods, N.   Fl. spring.   *A. trifòlia.*

45. CORNEL FAMILY. Order CORNACEÆ.

Sbrubs or trees (except our Dwarf Cornel), the calyx coherent with the ovary, which makes a berry-like stone-fruit; represented (except by the Tupelo or Pepperidge-tree, *Nyssa*, here omitted) only by the genus

1. Cornel (or Dogwood). *Cornus.*

Petals 4 and stamens 4, on the ovary.   Teeth of the calyx 4, very small.   Style 1.   Ovary 2-celled, in fruit berry-like with a 2-seeded stone.   Leaves entire, opposite, except in No. 7.   Flowers in spring or early summer.

✳ Flowers greenish, in a head, which is surrounded by a 4-leaved involucre resembling a large white corolla ; fruit bright red.

1. Dwarf Cornel (or Bunchberry).   Herb low, with 4 or 6 leaves near the top.   Damp woods.   *C. Canadénsis.*
2. Flowering C. or Dogwood.   Tree; leaves of the corolla-like involucre obcordate.   *C. flórida.*

    ✳ ✳ Flowers white, in flat and open cymes: shrubs

3. Round-leaved C.   Branches greenish. warty-dotted ; leaves round-oval, woolly beneath ; fruit pale blue.   Woods.   *C. circinàta.*
4. Silky C.   Branches purple ; young stalks and lower side of the ovate or oblong leaves silky woolly; fruit pale blue.   Swamps.   *C. sericea.*

385. Dwarf Cornel.   386. A separate flower enlarged.   387 A fruit cut across.

5. Red-Osier C.   Branches red-purple; leaves ovate, smooth, white and roughish beneath; fruit white.   Wet banks of streams.   *C. stonifera.*
6. Panicled C.   Branches gray; leaves lance-ovate; cymes convex; fruit white.   *C. paniculàta.*
7. Alternate-leaved C.   Branches greenish streaked with white; leaves crowded at the ends of the shoots, but alternate; leaves pointed; fruit bright blue.   Hill-sides.   *C. alternifòlia.*

- II. Monopetalous Division.

46. HONEYSUCKLE FAMILY. Order CAPRIFOLIACEÆ.

Shrubs or woody twiners (or one or two are herbs), distinguished by having a mono-petalous corolla bearing the 4 or 5 stamens, and borne on the ovary, and the leaves opposite without stipules.

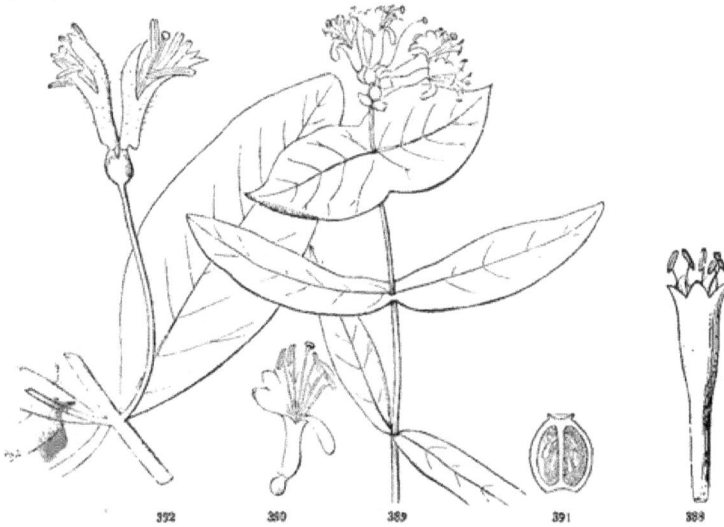

398. Flower of Trumpet-Honeysuckle. 388 Small-flowered Honeysuckle, 390. A separate flower, 391. An ovary divided lengthwise, and magnified. 392. Flowers, &c. of Fly-Honeysuckle, No. 11.

Herb creeping: the naked flower-stalk forking and bearing two sweet-scented, drooping, pretty flowers, with a 5-lobed and purple-tinged corolla hairy inside, but
    the stamens only 4,   (*Linnæa*) TWINFLOWER.
Shrubs or woody vines. Stamens as many as the lobes of the corolla, 4 or 5.
  Style one, slender: stigma one.
    Corolla elongated, mostly irregular. Berry several-seeded,   (*Lonicèra*) HONEYSUCKLE.
    Corolla elongated, nearly regular. Pod many-seeded,   (*Diervilla*) BUSH-HONEYSUCKLE.
    Corolla short bell-shaped, regular. Berry 2-seeded,   (*Symphoricàrpus*) SNOWBERRY.
  Style hardly any: stigmas generally 3: corolla very short and open, 5-cleft, regular.
    Flowers small, white, very many, in compound cymes.
    Leaves pinnate. Berry 3-seeded,   (*Sambucus*) ELDER.
    Leaves simple. Fruit berry-like with one flat stone,   (*Vibúrnum*) VIBURNUM.

Honeysuckle. *Lonicèra.*

Teeth of the calyx very short. Corolla tubular below, irregular and 2-lipped, four lobes belonging to one lip and one to the other, except in No. 1.

§ 1. Twining woody plants: flowers long, crowded in little heads at the end of the branches, or in sessile whorls in the axils of the uppermost leaves.

* Corolla long and narrow, appearing regular, the 5 short lobes nearly equal.

1. TRUMPET H. Uppermost pair of leaves united into one rounded body; corolla red, yellowish inside (also a yellow variety), scentless. Wild S. and cultivated.   *L. sempérvirens.*

* * Corolla 2-lipped: uppermost leaves on the flowering branches united round the stem into one flat or cup-shaped body, except in No. 2.

2. COMMON H. or WOODBINE. Leaves *all separate;* flowers purple-red outside, large, sweet-scented; berries red. Cultivated; as also the next.   *L. Periclýmenum.*

3. ITALIAN H. Leaves glaucous; flowers blush-colored, sweet-scented; berries yellow. *L. Caprifólium.*

4. WILD SWEET-H. Flowers smaller; otherwise nearly as in No. 3. S. and cultivated.   *L. grata.*

5. WILD YELLOW-H. Leaves thick, very glaucous both sides; several pairs united, flowers pale yellow; the tube rather long. W. and S.   *L. flava.*

6. SMALL-FL. H. Leaves glaucous; flowers small, yellowish and purplish or crimson. *L. parviflòra.*

7. HAIRY H. Leaves, &c. hairy, dull green, not glaucous; flowers clammy, orange. N.   *L. hirsùta.*

§ 2. Twining: leaves all separate, a pair of flowers in the axil of some of them, on a short 2-leaved footstalk. Cult. from Japan and China.

8. JAPAN H. Slender, hairy; corolla deeply 2-lipped, reddish outside, white inside, sweet. *L. Japónica.*

§ 3. Upright bushes: leaves all separate; flowers two on an axillary peduncle; their two ovaries often united at the base or into a double berry (Fig. 392): corolla short, irregular.

9. TARTARIAN H. Very smooth ; leaves somewhat heart-shaped; flowers rose-color, handsome, in spring. Cultivated for ornament.   *L. Tartárica.*

10. FLY H. Leaves petioled, ovate or heart-shaped, thin, a little hairy below and on the margins; corolla almost equally 5-lobed, greenish-yellow; ovaries separate. Woods, N.   *L. ciliàta.*

11. SWAMP FLY-H. Leaves sessile, oblong; peduncles long; corolla deeply 2-lipped, whitish. In swamps, N.   *L. oblongifolia.*

Elder. *Sambúcus.*

1. COMMON ELDER. Leaflets 7 to 11, smooth; cymes flat; berries dark purple.   *S. Canadénsis.*

2. RED-BERRIED E. Stems more woody; leaflets 5 or 7, downy beneath; cymes convex or pyramid-like; berries bright red. Cold woods, N.; fl. spring.   *S. pubens.*

Viburnum. *Vibúrnum.*

Shrubs or small trees, which have a variety of names. Leaves simple. Cymes flat. Fruit berry-like, with one flat stone. To the genus belongs the LAURESTINUS, cultivated in houses. All the following are wild in this country; but a variety of No. 6 is well known as a cultivated ornamental shrub. Flowering in spring or early summer.

* Flowers all alike, small and perfect: fruit blue or black.

1. NAKED V. or WYTHE-ROD. Leaves thickish, entire, or wavy-toothed. Swamps, N.   *V. nudum.*

2. SWEET V. or SHEEP-BERRY. Leaves ovate, pointed, very sharply serrate, on long and margined footstalks; cymes sessile; fruit rather large, eatable. A small tree. *V. Lentàgo.*

3. BLACK-HAW V. Leaves oval, blunt, shining; otherwise like No. 2. S. and W. *V. prunifòlium.*

4. ARROW-WOOD V. Leaves round-ovate, coarsely toothed, strongly marked with straight veins, smooth; cymes small, stalked; fruit small, bright blue. Shrub, in wet places. *V. dentàtum.*

5. MAPLE-LEAVED V. or DOCKMACKIE. Leaves roundish and with 3 pointed lobes, coarsely toothed, downy beneath; cymes long-stalked. Rocky woods: a shrub. *V. acerifòlium.*

* * Flowers at the margin of the cyme neutral, consisting merely of a large and flat corolla, white (just as in Hydrangea, p. 69, and Fig. 169.)

6. SNOWBALL V. or CRANBERRY-TREE. Leaves with 3 pointed lobes, smooth; fruit red, sour. Swamps, N. — The SNOWBALL-TREE or GUELDER-ROSE is a cultivated state of this, with all the flowers become neutral. *V. Ópulus.*

7. HOBBLEBUSH V. Branches long and spreading, often taking root; leaves large, round-ovate or heart-shaped, many-veined, scurfy beneath; cyme sessile, very broad; fruit red, turning blackish. Damp woods, N. *V. lantanoìdes.*

47. MADDER FAMILY. Order RUBIACEÆ.

Well distinguished by its regular monopetalous corolla, bearing 4 or 5 stamens alternate with its lobes, and itself borne on the ovary (the calyx being coherent); and the leaves in whorls, or else opposite and with stipules between them.

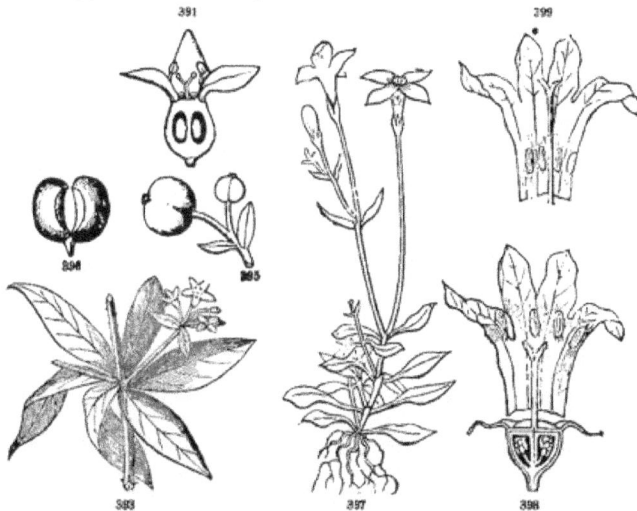

393. Piece of Madder, in flower. 394. Half of a flower, magnified. 395 Young fruits. 396. Ripe fruit. 397. Common Bluets 398 Section of a flower lengthwise, magnified, and the corolla laid open. 399. Corolla of another flower laid open, and the style.

1. Leaves in whorls. Ovary 2-celled, separating in the ripe fruit into two closed and one-seeded pieces: teeth or limb of the calyx small or hardly to be discerned.

Stamens 5 and the corolla 5-parted.   Fruit berry-like when ripe,   (*Rùbia*) *MADDER.
Stamens and divisions of the wheel-shaped corolla 4, rarely 3.   Fruit a pair of dry or
    fleshy akenes, smooth in some species, in others rough, in others beset with
    hooked prickles, making little burs,   (*Gàlium*) BEDSTRAW.

2. Leaves opposite, and with stipules, either as little scales or forming a small sheath.

Shrub: flowers (white) many in a close round head (Fig. 145),   (*Cephalànthus*) BUTTONBUSH.
Small herbs. (Corolla 4-lobed.)
    Flowers twin, on one ovary, which makes a double-eyed red berry. Small creeping
        evergreen, with round leaves. Corolla bearded inside. (*Mitchélla*) PARTRIDGE-BERRY.
    Flowers separate, peduncled. Fruit a dry pod. Stems erect. (*Oldenlàndia*, § *Houstònia*) BLUETS.

48. VALERIAN FAMILY.   Order VALERIANACEÆ.

Herbs, with strong-scented roots, opposite leaves, and no stipules, a 5-lobed monopetalous corolla bearing only 2 or 3 stamens, and borne on the ovary, which makes a small one-seeded dry fruit.   Flowers small, in cymes or clusters, white or purplish.

Limb of the calyx crowning the fruit in the form of feathery bristles,   (*Valeriàna*) *VALERIAN.
Limb of the calyx only one or more blunt teeth,   (*Fèdia*) LAMB-LETTUCE.

49. TEASEL FAMILY.   Order DIPSACEÆ.

Herbs, with opposite leaves, no stipules, and perfect flowers in dense heads, surrounded by an involucre, and with a chaffy bract under each blossom.   Corolla tubular or funnel-form, with 4 or 5 lobes, bearing 4 stamens, and itself borne on the ovary, which becomes an akene in fruit, containing one hanging seed.

Flowers in a rough-chaffy head: calyx cup-shaped, short: lobes of the corolla 4.   Stem
    and leaves rough or prickly,   (*Dipsacus*) TEASEL.
Flowers larger than the chaff: calyx with long-awned or bristle-shaped lobes: lobes of the
    corolla 4 or 5, unequal,   (*Scabiòsa*) *SCABIOUS.

50. COMPOSITE or SUNFLOWER FAMILY.   Order COMPOSITÆ.

Known by having what were called *compound flowers*, which are really a number of

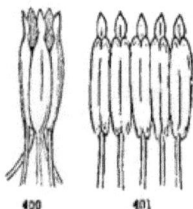

flowers closely crowded into a head, and this surrounded by an involucre which was taken for a calyx. The Scabious has its flowers in such heads.   But the distinguishing mark of the present family is that its five stamens are united by their anthers, or *syngenesious.*   Fig. 400 shows the stamens, their anthers connected into a tube, through which the style passes. Fig. 401 shows this tube split down on one side and spread open flat.   What gives the whole head so much the appearance of one large blossom is, that, in most cases, these flowers have a *strap-shaped* corolla. This will be understood by sup-posing a long tubular corolla to be split down on one side and spread out flat. In the Cichory (Fig. 402), Dandelion, and the like, all the flowers are strap-shaped. But in Sun-flower, Coreopsis (Fig. 404), Aster, and many others, only the flowers round the margin are strap-shaped; these are called *rays* or ray-flowers, and at first view much resemble the petals of a many-petalled blossom, — all the more so, be-cause in Coreopsis and Sun-flower these ray-flowers are *neutral*, having neither sta-mens nor pistils. But in As-ters and Daisies, they are *pis-tillate*, having a pistil only. The blossoms, which in these cases fill the body of the head, and are so small that the su-perficial observer is apt to take them for stamens or pis-tils, are regular and perfect, with a tubular and 5-lobed corolla (Fig. 405 *a*). They are called disk-flowers. In Thistles, Thoroughwort, Wormwood, and some kinds of Ground-sel, all the flowers are of this sort, i. e. there are no rays, but all the flowers tubular. In all, the ovary is one-celled and one seeded, and makes an akene in fruit. The corolla being on the ovary, the latter is of course covered by the

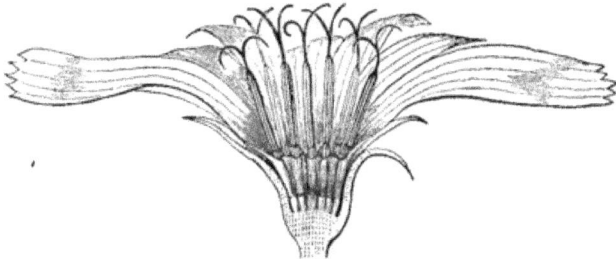

402. Flowers of Cichory, all with strap-shaped corollas.

403. Head of Cichory flowers, divided lengthwise and enlarged.

tube of the calyx adherent to it. Sometimes there is no limb or border to the calyx; then the akene is naked, as in that of Mayweed (Fig. 406). When the limb of the calyx is present in any form on the ovary or akene, it is named the *pappus* (which means seed-down). In Cichory the pappus or calyx is a ring or cup crowning the akene (Fig. 407); in Sunflower it consists of two chaffy scales, which fall off early (Fig. 408); in Helenium there are five chaffy and pointed scales (Fig. 409). But more commonly the pappus consists of bristles, or downy hairs (as its name denotes). Asters, Groundsels, and especially

Thistles, afford most familiar examples of such a hairy or downy pappus; those of Thistles, &c. in autumn sailing about in every breeze. Fig. 411 shows the very soft downy pappus of Sow-Thistle. Fig. 410, that of the Dandelion; this is raised upon a long beak to the akene, which lengthens greatly after flowering.

This family contains about an eighth or tenth part of all Flowering plants. But it is too difficult for the beginner. So we here barely mention a few of the common plants which belong to it.

404. Half of a head of flowers of Coreopsis.

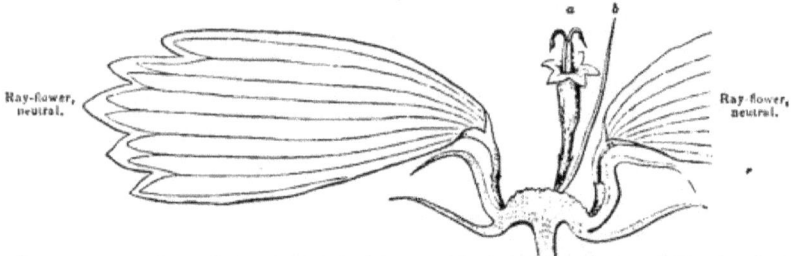

405. Slice of the same, enlarged, with one ray-flower, and part of another, and one perfect disk-flower (a), with its bract or chaff (b).

1. Among those which have no rays, or strap-shaped corollas, are *Thistles, Burdock, Everlasting* and *Cudweed, Wormwood, Thoroughwort* or *Eupatorium, Button-Snakeroot,* and *Ironweed.*

2. With rays or strap-shaped corollas at the margin (either neutral or pistillate), and tubular flowers in the centre; *Coltsfoot, Aster, Fleabane, Daisy, Golden-rod, Sunflower, Coreopsis, Mayweed, Chamomile, &c.*

3. With all the flowers strap-shaped and perfect (and

406   407   408   409   410   411

in this division the plants have a milky juice): *Cichory* or *Succory* (Fig. 402), *Salsify, Hawkweed, Sow-thistle, Dandelion,* and *Lettuce.*

51. LOBELIA FAMILY. Order LOBELIACEÆ.

Herbs with milky (acrid-poisonous) juice, alternate leaves, and scattered flowers, the stamens free from the peculiarly irregular corolla, which is split down on one side (Fig. 184), and borne with it on the many-seeded ovary. We have only one genus, viz. : —

Lobelia. *Lobèlia.*

Calyx with its short tube adherent to the 2-celled ovary, and with 5 slender teeth or lobes. Corolla unequally 5-lobed, and split down to the bottom on the upper side! Stamens 5, united into a tube both by their filaments and their anthers! Style one. Pod opening at the top. The following are the commonest wild species (all but Nos. 3 and 4 in low grounds); fl. summer and fall.

1. CARDINAL-FLOWER L. Tall, smooth, with a raceme of large, brilliant red flowers.   *L. cardinàlis.*

2. GREAT BLUE L. Rather hairy, 1° or 2° high; leaves lance-oblong; flowers 1' long, crowded in a leafy raceme, light blue.   *L. syphilitica.*

3. SPIKED L. Stem simple, straight, and slender, 1° to 3° high, including the long and naked spike-like raceme of small pale-blue flowers; lowest leaves obovate or oblong.   *L. spicàta.*

4. INDIAN-TOBACCO L. Branching, 8' to 18' high; leaves ovate-oblong; flowers very small, in irregular leafy racemes, pale blue; pods inflated. Open places.   *L. inflàta.*

52. CAMPANULA FAMILY. Order CAMPANULACEÆ.

Like the last family in all general respects, except that the showy corolla is regular, 5-lobed; the 5 stamens separate; the stigmas and the cells of the pod 3 or 5. Juice milky. The principal genus is

Campanula or Bellflower. *Campánula.*

So called from its generally campanulate or bell-shaped corolla (Fig. 179 and 412). The following are the commonest species.

   * Wild species: stigmas and cells of the pods 3.

1. HAREBELL C. A slender and very pretty plant, growing on shaded cliffs, 5' to 12' high; root-leaves round or heart-shaped, long-stalked, toothed; stem-leaves very narrow, entire; flowers nodding, the bright blue corolla bell-shaped, ½' or more long.   *C. rotundifòlia.*

2. MARSH C. A slender plant growing among grass, in wet places, with rough-angled stem and lance-shaped leaves; a few small pale flowers on diverging peduncles.   *C. aparinoìdes.*

3. TALL C. Stem tall, leafy, ending in a leafy loose spike (1° or 2° long) of blue flowers; corolla wheel-shaped; style long and curved. Rich low ground.   *C. Americàna.*

   * * Garden species: stigmas and cells of the pod 5.

412. Harebell.

4. CANTERBURY BELLS. Hairy, with stout stems, very large blue (or white) flowers, and broad appendages of the calyx covering the pod.   *C. Mèdium.*

53. HEATH FAMILY. Order ERICACEÆ.

Distinguished generally by the anthers opening by a pore or small hole at the top of each cell, and from all the other orders with a monopetalous corolla, except the two foregoing, by having the stamens free from the corolla, as many or twice as many as its lobes. But the petals are sometimes entirely separate, especially in the third and fourth sub-families. Fruit several-celled. Style one. This large order comprises four very distinct sub-families, viz. : —

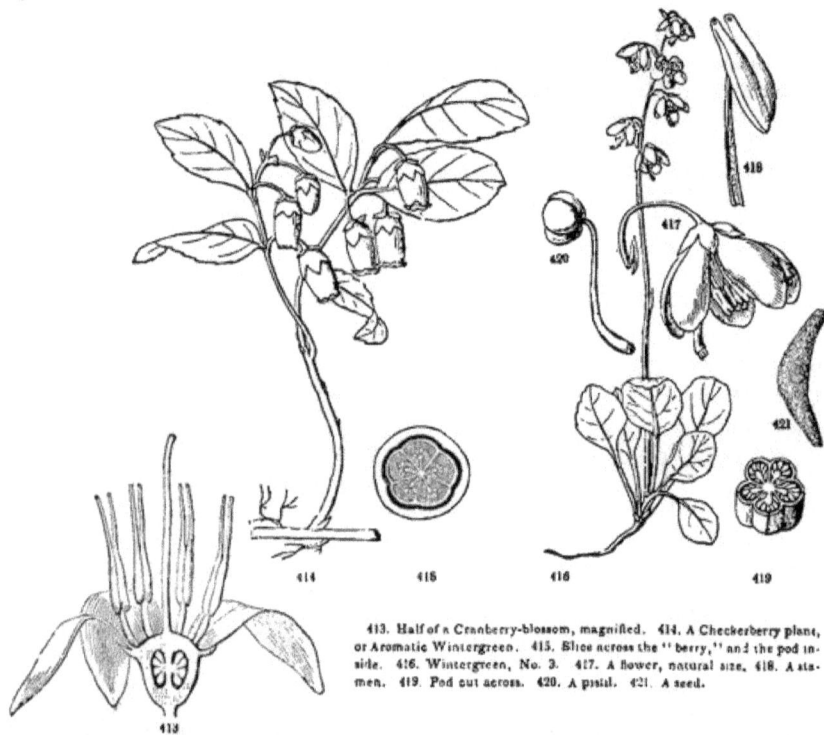

413. Half of a Cranberry-blossom, magnified. 414. A Checkerberry plant, or Aromatic Wintergreen. 415. Slice across the " berry," and the pod inside. 416. Wintergreen, No. 3. 417. A flower, natural size. 418. A stamen. 419. Pod cut across. 420. A pistil. 421. A seed.

I. HUCKLEBERRY Subfamily. Teeth of the calyx, corolla, and stamens on the ovary, the tube of the calyx coherent with its surface. Style and stigma one. Anthers of two nearly separate cells, tapering upwards into a tube or tip, which opens at the end. Shrubs, &c.

Ovary 10-celled with one ovule in each cell; berry with 10 largish seeds, or rather stones,
       in a circle,   (*Gaylussàcia*) HUCKLEBERRY.

Ovary with many ovules in each cell, making small seeds.
Stamens 10, rarely 8, included in the cylindrical or oblong-bell-shaped 5-toothed
  corolla. Berry blue or black, sweet, many-seeded,   (*Vaccinium*) BLUEBERRY.
Stamens 10, longer than the open bell-shaped 5-cleft corolla. Berry ripening few
  seeds, mawkish,   (*Vaccinium stamineum*) DEERBERRY.
Stamens 8, much projecting beyond the deeply 4-parted reflexed corolla. Berry
  4-celled, many-seeded, red, sour,   { *Vaccinium,* § *Oxycóccus*) CRANBERRY.

II. HEATH SUBFAMILY. Calyx, corolla (generally monopetalous), and stamens free from the ovary, inserted on the receptacle. Shrubby plants (except Checkerberry), sometimes small trees.
1. Corolla remaining dry after blossoming. Stems covered with very small and narrow
    leaves. Only house-plants in this country,   (*Erica*) *HEATH.
2. Corolla falling off after blossoming.
  Fruit a berry or berry-like.
    Trailing small-leaved evergreen. Corolla roundish,   (*Arctostáphylos*) BEARBERRY.
   Fruit a dry pod enclosed in a berry-like calyx,   (*Gaulthèria*) CHECKERBERRY.*
   Fruit a naked dry pod.
    Corolla salver-shaped, with a slender tube. A trailing, scarcely woody ever-
      green, with round-heart-shaped leaves,   (*Epigæa*) MAY-FLOWER.†
    Corolla ovate or oblong-cylindrical, 5-toothed,   (*Andrómeda*) ANDROMEDA.
    Corolla of 5 separate petals, regular, white.
     Flowers in panicled racemes, appearing in summer, (*Clethra*) SWEET-PEPPERBUSH.
     Flowers in umbels. Leaves rusty-woolly beneath,   (*Ledum*) LABRADOR-TEA.
     Flowers irregular, rose-purple, two of the petals nearly separate, (*Rhodòra*) RHODORA.
     Flowers bell-wheel-shaped, 5-lobed, with 10 pouches, (*Kálmia*) AMERICAN LAUREL.
     Flowers bell-shaped or short funnel-shaped without pouches, 5-lobed.
      Stamens 10. Leaves evergreen,   (*Rhododéndron*) RHODODENDRON.
      Stamens 5. Leaves falling in autumn,   (*Azàlea*) AZALEA.

III. WINTERGREEN or PYROLA SUBFAMILY. Calyx, &c. free from the ovary; the 5 separate petals and 10 stamens on the receptacle. Low and herbaceous, or nearly so, and with evergreen leaves.
Flowers in a raceme. Petals not widely spreading. Style long,   (*Pýrola*) WINTERGREEN.
Flowers in a general corymb or umbel, or only one or two. Style very short, (*Chimáphila*) PIPSISSEWA.

IV. INDIAN-PIPE SUBFAMILY. Low herbs growing in leaf-mould in woods, destitute of green foliage (parasitic on roots), having white or flesh-colored scales in place of leaves.
Flower one, nodding at first. Calyx of 2 to 4 scales : petals of 5 spatulate scales :
    stamens 10,   (*Monótropa*) INDIAN-PIPE.
Flowers several in a scaly raceme; the terminal blossom with 5 petals and 10 stamens,
    all the others with only 4 petals and 8 stamens,   (*Hypópitys*) PINESAP.

---

 * Called WINTERGREEN in the country in most places; also BOXBERRY or PARTRIDGE-BERRY; but the latter name rightly belongs to *Mitchella,* and that of Wintergreen to *Pyrola,* which is so named in England.
 † Also called TRAILING-ARBUTUS and GROUND-LAUREL. Nearly the earliest-flowering plant in the Northern States; prized for the rich spicy fragrance of its pretty rose-colored blossoms.

Huckleberry. *Gaylussàcia.*

Differing from Blueberries in the rather spicy and sweet berry having 10 large seeds, or rather small stones. The foliage and young shoots in the common species are sprinkled with waxy or sticky dots. Flowers purplish in racemes.

1. BLACK or COMMON H. Branches, leaves, &c. clammy when young; racemes and pedicels short; fruit black, without any bloom. Very common, furnishing the principal huckleberries of the market, ripe late in summer.   *G. resinòsa.*
2. PALE H. or BLUE-TANGLE. Leaves and fruit glaucous; pedicels long and drooping. *G. frondòsa.*
3. DWARF H. Branches rather hairy ; leaves thickish and shining ; racemes long, with leaf-like bracts. E. near the coast.   *G. dumòsa.*

Blueberry. *Vaccinium.*

Flowers white or tinged with pink, in short clusters, rather earlier than the leaves. Berries blue or black, and generally with a bloom, many-seeded. Leaves deciduous.

1. COMMON BLUEBERRY. Stem 5° to 10° high; leaves ovate, oval, or oblong. Swamps. *V. corymbòsum.*
2. LOW B. Stems 1° high, and obovate or oval glaucous leaves smooth.   *V. vacillans.*
3. DWARF B. Stems ½° to 1° high, smooth, leaves lance-oblong, fringed with fine bristle-pointed teeth, smooth, shining both sides. Dry woods, &c. This is the earliest *blueberry* or *blue huckleberry* in the market.   *V. Pennsylvànicum.*
4. CANADA B. Stems 1° or 2° high; branchlets and lance-oblong leaves downy: otherwise much like the last. N.   .   *V. Canadénse.*

Cranberry. *Vaccinium,* § *Oxycòccus.*

Slender, almost herbaceous, creeping or trailing, growing in bogs, with their small leaves rather crowded, entire, thickish, and evergreen, whitened beneath. Flowers single, nodding on the summit of a slender stalk, pale rose-colored, the corolla almost divided into 4 long and narrow petals turned back. Berries ripe in autumn.

1. LARGE CRANBERRY. Stems 1° to 3° long; leaves oblong, blunt, nearly flat, almost ½' long; berries ½' to 1' long, deep red (the principal *cranberry* of the market).   *V. macrocàrpon.*
2. SMALL CRANBERRY. Stems hardly 1° long; leaves ovate, acute, not half as large as those of No. 1, the margins more rolled back; berries much smaller, often speckled. N. and in mountain bogs.
                                                         V. Oxycòccus.

Kalmia or American Laurel. *Kàlmia.*

Flowers (in early summer) showy, in corymbs or umbels : an anther is at first lodged in each of the 10 pouches of the corolla. Leaves evergreen, very smooth.

1. MOUNTAIN L. or K. Leaves lance-ovate, bright green both sides; flowers large, pale or deep rose-color, in terminal corymbs; pedicels, &c. clammy. Stems 4° to 10° high.   *K. latifòlia.*
2. SHEEP L. or LAMBKILL. Leaves lance-oblong, blunt, pale beneath, petioled, mostly opposite, flowers small, purple; the corymbs becoming lateral; shrub 1° or 2° high.   *K. angustifòlia.*
3. PALE L. Leaves oblong, sessile, opposite, white-glaucous beneath; flowers few, large, lilac-purple. Swamps, N.   *K. glauca.*

Rhododendron (or Rose-Bay). *Rhododéndron.*

Calyx very small or obscure. Corolla large, 5-lobed. Stamens 10, more or less bent to one side, slender. Shrubs or low trees, with evergreen leaves and a corymb or umbel of large and handsome flowers from a terminal scaly bud, in early summer. We have only one common species, viz.: —

GREAT R. or LAUREL. Leaves lance-oblong, 4' to 10' long, green both sides; flowers 1' wide, pale rose or white, greenish, and spotted in the throat. Damp, deep woods. *R. máximum.*

Azalea. *Azàlea.*

Shrubs, like Rhododendron, but with thin and deciduous leaves; the long stamens only 5. Our two common wild species (wrongly called *Honeysuckle*) grow in swamps.

1. PURPLE A. or PINXTER-FLOWER. Flowers rather earlier than the leaves; corolla funnel-shaped with long recurved lobes, pink-purple or rose-color. *A. nudiflòra.*

2. CLAMMY or WHITE A. Flowers white, clammy, sweet-scented, later than the leaves, which are whitish or pale beneath. Common E. *A. viscòsa.*

Wintergreen (or Shin-leaf). *Pýrola.*

Leaves evergreen, rounded, all next the ground, around the base of a scape bearing a raceme of greenish-white (or rarely rose-colored) nodding flowers. Petals 5, all separate, not spreading. Stamens 10: filaments awl-shaped, naked. Style long. Pod 5-lobed.

* Style turned down and curved.

1. ROUND-LEAVED W. Leaves orbicular, thick, shining ; raceme many-flowered ; calyx-lobes lance-shaped. Moist woods. *P. rotundifòlia.*

2. OVAL-LEAVED W. Leaves broadly oval, thin; flowers many; calyx-lobes ovate, short. *P. elliptica.*

3. SMALL W. Leaves roundish, thick, small; flowers few; cells of the anther pointed. *P. chloràntha.*

* * Style straight.

4. ONE-SIDED W. Leaves thin, ovate; flowers small, all on one side of the raceme. *P. secùnda.*

Pipsissewa. *Chimáphila.*

Leaves evergreen, oblong or lance-shaped, toothed, crowded or scattered on short ascending stems, which bears at the summit from 1 to 7 fragrant flesh-colored flowers in a corymb or umbel. Petals orbicular, widely spreading. Stamens 10; their filaments enlarged and hairy in the middle. Style very short: stigma broad and flat. Dry woods; fl. early summer.

1. UMBELLED P. (or PRINCE'S-PINE). Leaves lance-shaped with a tapering base, serrate, bright green, not spotted; flowers 4 to 7. *C. umbellàta.*

2. SPOTTED P. Plant smaller, 3' to 5' high: leaves lance-ovate, obtuse at the base, blotched with white, flowers 1 to 4. *C. maculàta.*

54. HOLLY FAMILY. Order AQUIFOLIACEÆ.

Trees or shrubs, with alternate leaves, and small regular (often polygamous) flowers in the axils; the minute calyx and the 4–6-parted (greenish or white) corolla free from the ovary. Stamens 4 to 6, attached to the very base of the corolla, alternate with its divisions. Anthers opening lengthwise. Stigmas nearly sessile. Fruit a berry-like drupe, containing 4 to 6 small seedlike stones. — Consists mainly of the genus

Holly. *Ilex.*

Containing several species, some with deciduous, others with evergreen leaves.

1. AMERICAN HOLLY. Leaves thick and evergreen, spiny-toothed, oval ; parts of the blossom in fours; fruit red. — Tree with ash-colored bark and white wood. *I. opàca.*

2. Winterberry H. or Black Alder. Leaves thin and deciduous, serrate, veiny, ohovate or oblong; peduncles very short; parts of the blossom often in sixes; fruit red.   Shrub: low grounds.
This belongs to the section *Prinos*.   *I. verticillàta.*

55. EBONY FAMILY.   Order EBENACEÆ.

Of this small family, we have only one species, a tree, which deserves notice, viz. : —

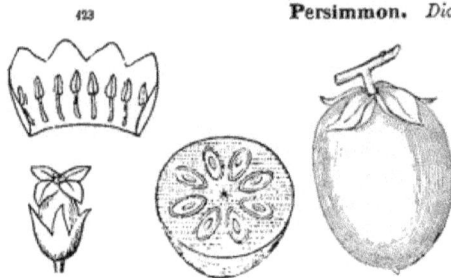

Persimmon. *Diospýros.*

Tree with alternate thickish leaves;
in their axils some trees bear clustered
staminate flowers, with a 4-cleft corolla
and about 16 stamens; others single and
larger perfect flowers, with a 4-lobed
corolla and 8 stamens. Calyx 4-cleft,
rather large, thickish. Corolla pale yellow. Pistil one, with 4 styles: the ovary
ripening into a plum-like fruit, which is
very astringent when green, but sweet
and yellow and eatable after frosts, containing 8 large and bony flat seeds.
                                                                 D. Virginiàna.

422. Fertile flower. 423. Corolla and stamens of the same, laid open.
424. Fruit. 425. Section of the same.

56. PLANTAIN FAMILY.   Order PLANTAGINACEÆ.

Consists mainly of the genus of low stemless herbs called

Plantain (or Rib-Grass). *Plantàgo.*

Flowers greenish, on a scape, in a close spike. — Calyx of 4 persistent sepals. Corolla
salver-shaped, thin, withering on the pod, 4-lobed. Stamens 4, generally with very long
and weak filaments, borne on the corolla. Style and stigma one, slender. Pod 2-celled,
opening crosswise, the top falling off as a lid, the loose partition falling out with the
seeds. Leaves generally with strong ribs.

1. Common Plantain. Leaves ovate or
slightly heart-shaped, several-ribbed;
seeds 7 to 16.   *P. major.*

2. Virginia P. Small (2' to 7' high), hairy;
leaves oblong, 3–5-ribbed; seeds 2.
                                P. Virginica.

3. English P. or Ripple-Grass. Hairy,
with long lance-shaped or linear leaves,
and a short and thick spike or head, on
a scape 1° or 2° high; seeds 2. Common E.   *P. lanceolàta.*

4. Seaside P. Smooth; leaves linear, thick
and fleshy; seeds 2. Salt marshes on the
coast.   *P. maritima.*

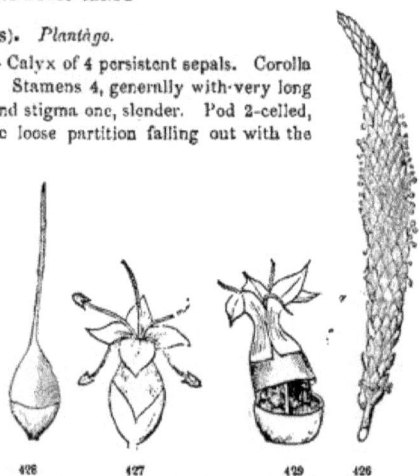

426. Young spike of common Plantain. 427. A flower magnified. 428. Pistil of the same. 429. Fruit, opening by a lid, the withered corolla on the
lid.

57. LEADWORT FAMILY. Order PLUMBAGINACEÆ.

Familiar to us in two plants only, viz. MARSH-ROSEMARY on the coast, and THRIFT in gardens; known by having a dry and scaly funnel-shaped calyx, and 5 petals united only at their base, with a stamen before each, and 5 styles on a single one-seeded ovary.

Flowers (rose-color) in a round head on a long and naked scape: leaves very narrow, all in a close tuft at the root,   (*Armèria*) THRIFT.

Flowers (lavender-color) spiked or sessile along the branches of a forking panicle : leaves spatulate, thickish, on petioles, nearly all of them from the stout rootstock,   (*Stàtice*) MARSH-ROSEMARY.

430. Calyx and corolla of Thrift ; separated. 431. Pistil of the same, with its 5 styles : also the lower part of an ovary more magnified, cut across.

430   431

58. PRIMROSE FAMILY. Order PRIMULACEÆ.

Herbs, with regular perfect flowers; completely distinguished by having the stamens of the same number as the lobes to the corolla and *one before each*, inserted on the tube ; the pistil with a one-celled ovary or pod, with one large placenta rising from its base, and bearing many or few seeds.

Leaves under water pinnately divided into thread-like divisions; flowering stems hollow, and inflated between the joints,   (*Hottònia*) FEATHERFOIL.

Leaves simple and entire or barely toothed.

    Calyx with its tube coherent with the base of the ovary. Flowers very small, white, in racemes. Leaves alternate,   (*Sàmolus*) BROOKWEED.

    Calyx and corolla free, inserted on the receptacle.

      Leaves all at the root: flowers in an umbel.

        Calyx tubular: corolla salver-shaped: stamens included,   (*Primula*) PRIMROSE.

        Calyx and corolla 5-parted, turned back: anthers long and filaments very short, connected,   (*Dodecàtheon*) DODECATHEON.

      Leaves several in a whorl at the summit of the slender stem. Calyx and corolla 7-parted, wheel-shaped, with narrow divisions,   (*Trientàlis*) STAR-FLOWER.

      Leaves (mostly opposite or whorled) borne along the whole length of the stem: corolla 5-parted.

        Corolla wheel-shaped, yellow,   (*Lysimàchia*) LOOSESTRIFE.

        Corolla wheel-shaped, blue or purple: pod opening by a lid,   (*Anagàllis*) PIMPERNEL.

Loosestrife. *Lysimàchia.*

This is the only genus in the Primrose family of which we have more than one common wild species. The 5 stamens have their filaments a little monadelphous at the base, and often unequal. Fl. in summer.

1. STRICT L. Leaves opposite or scattered, lance-shaped: stem ending in a long raceme leafy at the base; divisions of the corolla lance-oblong. Low grounds.   *L. stricta.*

2. **Four-leaved L.** Stem simple; leaves lance-ovate, in whorls of 4 (sometimes of 3 or 6); flowers long-stalked from the axil of the leaves. Sandy grounds.   *L. quadrifòlia.*

8. **Ciliate L.** Leaves opposite, lance-ovate, with a rounded or heart-shaped base, on long ciliate footstalks; flowers long-stalked from the upper axils; divisions of the corolla ovate, pointed, and with wavy or slightly toothed margins. Low grounds.   *L. ciliàta.*

4. **Lance-leaved L.** Leaves lance-shaped, oblong, or linear, narrowed into a short margined footstalk; flowers, &c. nearly as in No. 3. S. & W. in low grounds.   *L. lanceolàta.*

59. BIGNONIA FAMILY. Order BIGNONIACEÆ.

Plants with mostly opposite leaves, and large and showy flowers: the corolla 2-lipped or rather irregular, bearing on its tube 4 stamens (2 long and 2 short) or only 2, often with rudiments of the other one or three. Fruit a large 2-celled pod, with many large seeds: the whole kernel is a flat embryo. Calyx free and corolla on the receptacle, as it is in all the following families with monopetalous corolla.

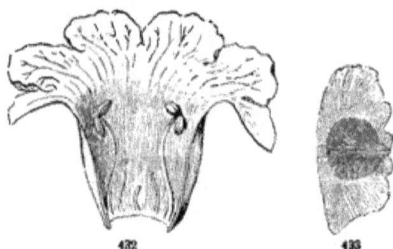

432. Corolla of Catalpa laid open, with the stamens.
433. Winged seed of Trumpet-Creeper.

Woody plants, with winged seeds, in long pods. Vine climbing by rootlets: leaves pinnate. Calyx 5-toothed. Corolla funnel-shaped, 5-lobed: stamens 4,   (*Tecoma*) Trumpet-Creeper.

Tree, with simple heart-shaped leaves, and white flowers (purple-tinged or dotted) in large panicles. Calyx 2-lipped. Corolla bell-shaped and 2-lipped: stamens generally 2, with vestiges of one or three others,   (*Catàlpa*) Catalpa.

Rank clammy herb (cult. and wild S. W.) with wingless seeds in a large and long-pointed fruit, the outer part of which is fleshy and falls off from the inner fibrous-woody part: this is crested and long-beaked, the beak at length splitting into 2 hooked horns. Corolla dull-colored: stamens 2 or 4, perfect,   (*Martýnia*) Unicorn-plant.

60. BROOM-RAPE FAMILY. Order OROBANCHACEÆ.

Herbs parasitic on the roots of trees, &c., readily known by their irregular monopetalous corolla, 4 stamens, in two pairs; the ovary one-celled with innumerable small seeds on the walls. Also, like other parasitic plants, they are entirely destitute of green herbage, yellowish or brownish throughout, and with scales in place of leaves.

Stems slender and branched, with few and small scales and many flowers scattered along the branches,   .   (*Epiphègus*) Beech-drops.

Stems short and thick, covered with broad scales, so that the plant resembles a fir-cone Flowers under the upper scales: stamens projecting,   (*Conóphulis*) Squaw-root.

Stems or naked and 1-flowered scapes slender, from a scaly base: stamens included in the curved and salver-shaped corolla,   (*Aphýllon*) Naked Broom-rape.

61. FIGWORT FAMILY. Order SCROPHULARIACEÆ.

Herbs with a 2-lipped or more or less irregular monopetalous corolla, and 4 stamens in pairs (2 long and 2 short), or only 2 perfect stamens; rarely all 5 present: style 1: the ovary 2-celled and making a many-seeded pod (few-seeded in some Speedwells and Cow-wheat). Flowers often showy. Two lobes always belong to the upper lip, three to the lower.

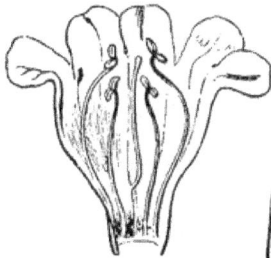

434. Corolla of a Pentstemon, laid open, showing the 4 perfect stamens and the fifth as a sterile naked filament.

435. Stamens (with a piece of the corolla) of another Pentstemon, with the sterile filament bearded.

436. Piece of Purple Gerardia.

437. Corolla laid open, showing the 4 stamens in pairs.

438. Style and calyx of the same.

439. Part of a pod.

440. Flower of Toadflax.

441. Plant of Hedge-Hyssop.

442. Flower laid open; one pair of good stamens; one pair of sterile filaments.

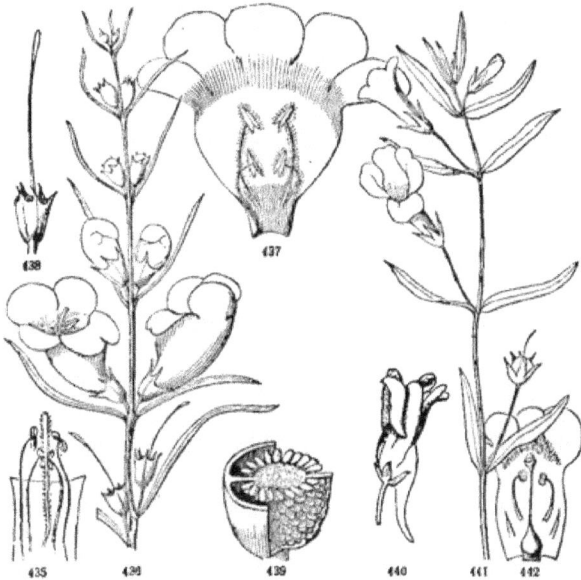

* Corolla wheel-shaped or with a very short tube, the lobes more or less unequal.

Calyx and corolla 5-cleft: stamens 5, some of them rather imperfect, (*Verbáscum*) MULLEIN.

Calyx and corolla 4-parted: stamens 2, (*Verónica*) SPEEDWELL.

* * Corolla more or less tubular, bell-shaped and irregular, or 2-lipped.

← Upper lip or lobes covering the lower in the bud (except sometimes in Monkey-flower).

Corolla with its 2-lipped mouth closed by a palate, i. e. an inward projection of the lower lip: stamens 4.

Corolla with a slender spur at the base on the lower side, (*Linária*) TOADFLAX.

Corolla sac-like at the base on the lower side, (*Antirrhinum*) SNAPDRAGON.

Corolla 2-parted: the lower lip sac-shaped in the middle; the short tube with a protuberance at the base on the upper side: stamens 4, (*Collinsia*) COLLINSIA.

Corolla ovoid, small, dull greenish purple, with 4 short unequal erect lobes, and one small
         recurved one (the lower). Stamens 4 and a rudiment, (*Scrophulària*) FIGWORT
Corolla shaped like a turtle's head, the mouth closed or nearly so, without a palate. Sta-
         mens 4 with woolly anthers; and a sterile filament besides, (*Chelòne*) TURTLEHEAD.
Corolla open at the irregular or 2-lipped mouth. Stamens 4, and a sterile filament besides
         (Fig. 434, 435),   (*Pentstèmon*) PENTSTEMON.
Corolla. 2-lipped; the upper lip with the sides turned back, the lower lip turned down.
         Stamens 4, no vestige of the fifth. Calyx elongated, 5-angled, 5-
         toothed. Stigmas with 2 broad lips,   (*Mimulus*) MONKEY-FLOWER.
Corolla somewhat 2-lipped, open. Stamens only 2 perfect. Calyx 5-parted.
    Sterile filaments included, or none. Corolla yellow or whitish,   (*Gratìola*) HEDGE-HYSSOP.
    Sterile filaments long, protruding from the purple or blue corolla, (*Ilysánthes*) FALSE-PIMPERNEL.

    + + Lower lip or the side lobes covering the others in the bud.

Corolla (large, purple or white) tubular, open; the border slightly 5-lobed, (*Digitàlis*) *FOXGLOVE.
Corolla salver-shaped. Flowers in a spike.
    Stamens 2, projecting, longer than the 4 lobes of the corolla, (*Verónica Virgínica*) CULVER'S-ROOT.
    Stamens 4, included : lobes of the corolla 5 : calyx tubular, 5-toothed, (*Búchnera*) BLUE-HEARTS.
Corolla bell-shaped or funnel-shaped, somewhat irregularly 5-lobed. Stamens 4, (*Gerárdia*) GERARDIA.
Corolla tubular, decidedly 2-lipped, the narrow upper lip erect or arched, enclosing the 4
         stamens. Flowers in a spike. Pod many-seeded.
    Bracts large and colored, scarlet in our species. Calyx tubular,   (*Costilleia*) PAINTED-CUP.
    Bracts green, small. Leaves pinnatifid,   (*Pediculàris*) LOUSEWORT.

Mullein. *Verbáscum.*

   Flowers in a long terminal spike or raceme. Corolla 5-parted, almost regular. Stamens 5, unequal,
but generally all with anthers. Root biennial.

1. COMMON M. Tall, woolly throughout; the simple stem winged by the prolonged bases of the leaves;
   flowers yellow, in a long thick spike; two of the filaments smooth. Fields, &c.   *V. Thapsus.*
2. MOTH M. Green, smoothish; stem 2° or 3° high; leaves toothed; flowers yellow or white in a loose
   raceme; filaments all bearded with yellow wool. Road-sides.   *V. Blattària.*

Speedwell. *Verónica.*

   Flowers small; one or two of the lobes of the 4-parted border of the corolla always smaller than the
others. Stamens 2, protruding. Pod flattened, many-seeded in the common species.

   § 1. Corolla salver-shaped, the tube longer than the border. Pod not notched at the end.

1. CULVER'S-ROOT S. A tall perennial, with lance-shaped pointed leaves in whorls, and whitish flowers
   crowded in clustered spikes. Woods, W. and S., and cultivated in gardens.   *V. Virgínica.*

   § 2. Corolla wheel-shaped, tube very short, pale blue or white. Pod notched at the end. Leaves opposite.

         * Flowers in single racemes from the axils of the leaves.

2. WATER S. Smooth; stems rooting at the creeping lower part, then erect; leaves sessile by a heart-
   shaped base, ovate-lanceolate; corolla pale blue with darker stripes. Brooks.   *V. Anagóllis.*
3. BROOK S. or BROOKLIME. Leaves ovate or oblong, on petioles; otherwise like the last. *V. Americàna.*
4. MARSH S. Smooth, slender; leaves sessile, linear, acute; raceme zigzag, loose.   *V. scutellàta.*
5. COMMON S. Downy; stems creeping; leaves wedge-oblong, serrate; raceme dense.   Dry ground,
   in open woods.   *V. officinàlis.*

* * Flowers in a terminal loose raceme.

6. THYME-LEAVED S. Smooth and small, 2' to 4' high from a creeping base; leaves ovate or oblong, the lowest petioled and rounded. Fields, everywhere. *V. serpyllifòlia.*

* * * Flowers in the axils of the upper leaves. Root annual.

7. PURSLANE S. or NECKWEED. Smooth, branching, erect; lower leaves oval or oblong, toothed, and petioled; uppermost oblong-linear, sessile, and entire. Cult. grounds, &c. *V. peregrìna.*

8. CORN S. Hairy; lower leaves ovate, crenate, petioled; the upper sessile, lance-shaped, and entire. Cultivated grounds. *V. arvénsis.*

Toadflax. *Linària.*

1. COMMON T. (BUTTER-AND-EGGS, RAMSTED). Stems branching, crowded with the pale linear leaves; flowers crowded in a close raceme, large and showy, pale yellow with the palate orange-colored. A weed in fields and road-sides. *L. vulgàris,*

2. WILD T. Stem very slender, simple, with scattered linear leaves; prostrate shoots at the bottom with broader leaves; flowers very small, blue, in a slender raceme. Sandy soil. *L. Canadénsis.*

Gerardia. *Gerárdia.*

Plants with large and showy somewhat leafy-racemed flowers; the corolla a little irregular, but hardly 2-lipped. Stamens woolly or hairy; the 4 anthers approaching in pairs. Fl. late summer and autumn.

* Corolla rose-purple: calyx bell-shaped, with 5 short teeth: plants low and busby-branched.

PURPLE G. Leaves linear, rough-margined; flowers 1' long, short-stalked. *G. purpùrea.*

SLENDER G. Leaves linear; flower about ½' long, on a long and slender stalk. *G. tenuifòlia.*

* * Corolla yellow, with a rather long tube, woolly inside: calyx 5-cleft, leaf-like.

3. DOWNY G. . Stem (3° or 4° high) and oblong or lance-shaped leaves clothed with a fine close down, upper leaves entire, lower ones sinuate or pinnatifid. Woods. *G. flava.*

4. SMOOTH G. Smooth throughout and glaucous, 3° to 6° high; lower leaves twice pinnatifid, upper once pinnatifid or entire. Rich woods. *G. quercifòlia.*

5. CUT-LEAVED G. rather downy, bushy-branched, 2° or 3° high, very leafy; leaves pinnatifid, the crowded divisions cut and toothed. *G. pediculària.*

62. VERVAIN FAMILY. Order VERBENACEÆ.

Herbs or shrubby plants, with opposite leaves, a 2-lipped or unequally 5- (or rarely 4-) lobed corolla, and 4 stamens in pairs (i. e. 2 long and 2 short ones): the pistil with a single ovary and only one seed in each cell; the fruit either berry-like with 4 stones, or dry and splitting into 2 or 4 akenes, or in Lopseed consisting of a single akene. This family is intermediate between the foregoing order and the next. The two following are the commonest genera.

Calyx cylindrical, 2-lipped. Corolla 2-lipped. Ovary 1-celled, simple. Herb, in woods, with small whitish flowers in slender and loose spikes; the calyx containing the akene, turned down in fruit, (*Phryma*) LOPSEED.

Calyx tubular, 5-toothed. Corolla salver-shaped, with 5 slightly unequal lobes. Flowers in spikes or heads, summer and autumn, (*Verbèna*) VERVAIN.

Vervain. *Verbèna.*

* Showy Verbenas: low and showy-flowered species, in gardens in summer, the greater part from South America, viz. *V. Melíndres* (red) and others, now much mixed. And there is one species of this sort wild in Western prairies, viz.: —

1. Aublet's Verbena. Rather hairy; leaves pinnatifid or cut; spikes flat-topped in blossom, like a corymb; corolla light purple, &c.   *V. Aublètia.*

* * Common Vervains: weeds or weed-like plants, in fields and road-sides, with small flowers in long spikes, which are generally panicled.

2. Common V. Erect, slenderly branched, 1° to 3° high; leaves sessile, cleft or pinnatifid and cut-toothed; spikes very slender; flowers very small, purplish.   *V. officinàlis.*

3. White V. Leaves petioled, ovate or oval, serrate; spikes of white flowers very slender.   *V. urticifòlia.*

4. Blue V. Leaves petioled, lance-shaped or lance-oblong, the lower often cut or 2-lobed at the base; spikes of blue flowers thick and close; stem 4° to 6° high.   *V. hastàta.*

5. Low V. Stems ½° to 1° high; leaves lance-linear, sessile, scarcely toothed; spikes one or few, thickish; flowers purple. S. and W.   *V. angustifòlia.*

63. SAGE or MINT FAMILY.   Order LABIATÆ.

Herbs with square stems and opposite aromatic leaves, a 2-lipped (or rather irregular)

corolla, 4 stamens in pairs (2 long and 2 shorter), or else only 2 stamens, and a 4-parted ovary, in fruit making 4 akenes around the base of the single style. That is, among the families with 2-lipped or irregular monopetalous corollas this is at once known by the 4-lobed ovary, making 4 akenes. The leaves are commonly more or less dotted with small glands, which contain a volatile oil, peculiar to each species. This gives the warm aromatic properties which all plants of this family possess. By distillation, the oil is extracted from several species, as from Peppermint and Spearmint, Lavender, Pennyroyal, &c. Or the dried foliage is used for seasoning or for herb drinks in the case of Summer-Savory, Marjoram, Thyme, Catnip, and Sage. The following are the common genera or kinds of this large family.

443   Flower of Garden Sage. 444. Pistil of the same, the 4-lobed ovary in the bottom of the calyx, half of which is cut away.

* Stamens 4, turned down so as to rest upon the lower lip of the corolla.

Flowers in racemes, white: calyx soon reflexed, its upper lobe large and round: upper
     lip of the corolla 4-cleft, the lower entire. Leaves ovate, fragrant, (*Ócimum*) *Sweet-Basil.

Flowers in a naked and peduncled spike, pale blue: calyx narrow, 5-toothed: the 5 lobes
     of the corolla almost equal: stamens short: leaves narrow, hoary, (*Lavándula*) *Lavender.

* * Stamens 4, ascending, and projecting from the upper side of the corolla. Akenes veiny.

Corolla cleft down the upper side, the lower lobe much larger than the other 4. Flowers
     purplish, rarely white, in a spike,   (*Teùcrium*) Germander.

Corolla with the border cleft into 5 almost equal lobes, blue.

Stamens very long, curved: lobes of the corolla turned rather forward, (*Trichostéma*) Blue-curls.

Stamens slightly projecting from the equally 5-lobed corolla,   (*Isánthus*) False Pennyroyal.

Stamens 4 or 2, not turned down, and not protruding from the upper side of the flower.
Corolla scarcely at all two-lipped, almost equally 4-lobed. Flowers small.
  Stamens 4 with anthers, almost equal in length,   (*Mentha*) MINT.
  Stamens only 2 with anthers. Flowers in dense axillary whorls, (*Lycopus*) WATER-HOREHOUND.
Corolla evidently 2-lipped: stamens 2, or only 2 with anthers.
  Upper lip nearly flat or spreading, 2-lobed or notched at the end.
    Calyx equally 5-toothed, bearded in the throat. Cymes terminal,   (*Cunila*) DITTANY.
    Calyx 2-lipped: upper lip 3-toothed, the lower 2-cleft.
      Throat of the calyx bearded: corolla small: 2 sterile filaments, (*Hedeoma*) PENNYROYAL.
      Throat of the calyx naked; that of the large corolla bearded; the middle
          lobe of its lower lip large and hanging, fringe-toothed, (*Collinsonia*) HORSE-BALM.
  Upper lip of the corolla arched, entire or slightly notched, holding the stamens.
    Calyx equally 5-toothed, tubular: lips of the large corolla long and narrow.
      Flowers crowded in close and leafy-bracted heads,   (*Monarda*) HORSE-MINT.
    Calyx 2-lipped.'
      Upper lip with 3 bristle-pointed teeth. Flowers in heads,   (*Blephilia*) BLEPHILIA.
      Upper lip entire or 3-toothed. Anthers with only one cell, on the end of a
          long connective astride the end of the filament,   (*Salvia*) SAGE.
Corolla 2-lipped: stamens 4, all with anthers.
  Upper and inner pair of stamens longer than the lower or outer pair,
    And curved downwards. Flowers spiked, small. Herbs tall, (*Lophanthus*) GIANT-HYSSOP.
  Both pairs of stamens ascending under the upper lip.
    Flowers in terminal spikes or clusters,   (*Nepeta*) CATNIP.
    Flowers few in the axils of kidney-shaped leaves,   (*Glechoma*) GROUND-IVY.
  Upper pair of stamens shorter than the lower or outer pair.
    Upper lip of the corolla flat and open, or barely concave.
      Stamens distant or diverging, not approaching under the upper lip.
        Calyx tubular, equally 5-toothed, 15-nerved. Stamens long,   (*Hyssopus*) *HYSSOP.
        Calyx 10 to 13-nerved, ovate, bell-shaped, or short tubular.
          Calyx naked in the throat.
            Flowers in dense heads or clusters,   (*Pycnanthemum*) MOUNTAIN-MINT.
            Flowers clustered in the axils or spiked,   (*Satureia*) *SUMMER-SAVORY.
          Calyx hairy in the throat.
            Flowers spiked, and with large colored bracts,   (*Origanum*) MARJORAM.
            Flowers loosely clustered: bracts minute,   (*Thymus*) *THYME.
      Stamens with their anthers approaching in pairs under the upper lip.
        Calyx tubular. Flowers in a head-like cluster, surrounded with awl-
          shaped bracts,   (*Clinopodium*) BASIL.
        Calyx tubular-bell-shaped and 2-lipped: corolla curved upwards.
          Flowers in loose clusters,   (*Melissa*) *BALM.
    Upper lip of the corolla concave, the whole throat inflated and funnel-shaped.
        Flowers large in naked spikes,   (*Physostegia*) FALSE-DRAGONHEAD.
  Upper lip of the corolla arched or hood-like.
    Calyx 2-lipped, closed over the fruit, and
      Very veiny, the lips toothed: flowers in a bracted short spike,   (*Brunella*) SELF-HEAL.
      Not veiny, becoming helmet-shaped; lips entire,   Scutellaria) SCULLCAP.

Calyx not 2-lipped, 10-toothed. Clusters axillary, head-like,   (*Marrúbium*) HOREHOUND.
Calyx not 2-lipped and only 5-toothed,
    Funnel-shaped and much larger than the corolla,   (*Moluccélla*) *MOLUCCA-BALM.
    Bell-shaped or top-shaped, much shorter than the corolla.
      Anthers opening crosswise: calyx-teeth spiny-pointed,   (*Galeópsis*) HEMP-NETTLE.
      Anthers opening lengthwise.
        Corolla not enlarged in the throat: stamens turned down after shed-
            ding their pollen,   (*Stachys*) HEDGE-NETTLE.
        Corolla enlarged in the throat: calyx-teeth not spiny, (*Làmium*) DEAD-NETTLE.
        Corolla not enlarged in the throat: calyx top-shaped with spiny
            teeth. Akenes 3-angled. Leaves cleft and cut, (*Leonùrus*) MOTHERWORT.

Mint. *Mentha.*

Herbs with sharp-tasted leaves and small whitish or purplish flowers: upper lobe of the short corolla either entire or notched.

1. WILD MINT. Flowers in head-like clusters around the stem in the axils of the petioled leaves;
    plant hairy, or in one variety smoothish. Wet places.   *M. Canadénsis.*
2. PEPPERMINT. Smooth; clusters of flowers crowded in short spikes; leaves petioled, oblong or
    ovate.   *M. pipérita.*
3. SPEARMINT. Nearly smooth; spikes panicled; leaves lance-ovate, almost sessile.   *M. víridis.*

Horse-Mint. *Monárda.*

Herbs with mostly simple stems, and rather large flowers in close head-like clusters at the summit
of the stem, and around it in the axils of the upper leaves, surrounded by large bracts.

    * Root perennial: upper lip of the narrow corolla entire, the 2 stamens projecting from it: leaves
            lance-ovate or slightly heart-shaped.

1. BALM H. or OSWEGO TEA. Green, rather hairy; corolla long, bright red; uppermost leaves and
    bracts tinged with red. Moist banks, N., and in gardens.   *M. dídyma.*
2. COMMON H. Pale, smoothish or soft downy; flowers purplish or whitish, smaller.   *M. fistulòsa.*

      * * Root annual: upper lip of the corolla notched: stamens not projecting.

3. DOTTED H. Leaves lance-shaped; bracts yellowish and purple; corolla yellowish, purple-spotted.
    Sandy soil, S.   *M. punctàta.*

Scullcap. *Scutellària.*

Well marked by the tubular ascending corolla (mostly blue or bluish-purple) with a strongly arched
upper lip; the calyx with two short entire lips, closed after the corolla falls, and having an enlargement
on the back, the whole becoming of the shape of a helmet. Fl. summer.

      * Flowers small, in axillary one-sided racemes.

1. MAD-DOG S. Smooth, branched, slender; leaves lance-ovate or oblong, pointed, serrate, on slender
    stalks. Wet places.   *S. lateriflòra.*
      * * Flowers in terminal racemes.
2. LARGER S. Hairy and rather clammy, 1° to 3° high; leaves heart-shaped or ovate, wrinkled-
    veiny; upper lip of the corolla blue, the lower pale and purple-spotted. S. and W.   *S. versícolor.*
3. HAIRY S. Hairy, 1° to 3° high, slender; leaves ovate, crenate, obtuse, veiny.   *S. pilòsa.*

4. NARROW-LEAVED S. Minutely hoary or downy, slender, 1° or 2° high; leaves lance-oblong or linear, entire; raceme short, as in the foregoing. E. and S. *S. integrifolia.*

    * * * Flowers single, in the axils of the leaves.

5. DWARF S. Minutely downy, 3′ to 6′ high; leaves round-ovate or the upper lance-ovate, entire, ½′ long. Dry or sandy banks of rivers, &c. *S. párvula.*

6. SLENDER S. Slender, 1° or 2° high; leaves lance-ovate, serrate, with a roundish or slightly heart-shaped base, sessile; flowers ⅜′ long. Wet woods. *S. galericulàta.*

64. BORRAGE FAMILY. Order BORRAGINACEÆ.

Herbs with alternate entire leaves, not aromatic, commonly rough: the flowers regular, with a 5-leaved calyx, 5-lobed corolla, 5 stamens on the tube, one style, and a 4-lobed ovary, making 4 akenes. Flowers generally in one-sided raceme-like clusters, coiled up at the tip, and unfolding as the blossoms expand. Innocent mucilaginous and slightly bitter plants, the roots of some species yielding a red dye.

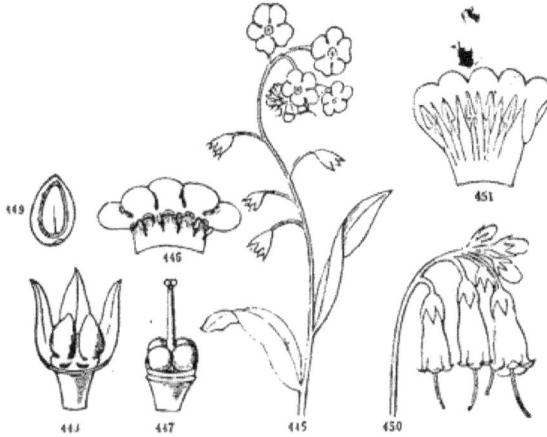

445. Branch of Forget-me-not, in flower.
446. The corolla laid open, with the stamens, magnified
447. The pistil with its 4-lobed ovary; calyx, &c. cut away.
448. Two of the ripe akenes in the calyx; the two sepals towards the eye and two of the akenes removed.
449. Akene cut through lengthwise, magnified; the whole kernel embryo.
450. Flowers of Comfrey.
451. Corolla enlarged, laid open, showing the sharp scales inside, and the stamens.

    * Ovary 4-parted, making 4 akenes around the base of the style.

Akenes or lobes erect, fixed by the lower end, separate from the style, not prickly.

    Corolla somewhat irregular (the lobes rather unequal), funnel-shaped (blue or purple).

        Its throat naked and open: stamens protruding, rather unequal, (*Échium*) VIPER'S-BUGLOSS.

        Its throat closed by 5 blunt scales; tube curved: stamens included, (*Lycópsis*) BUGLOSS.

    Corolla, &c. perfectly regular.

        Its throat closed by 5 converging scales, one before each lobe.

            Corolla wheel-shaped; its lobes acute. Plant rough-bristly, (*Borràgo*) *BORRAGE.

            Corolla tubular and somewhat funnel-shaped, 5-toothed, (*Sýmphytum*) COMFREY.

        Its throat open, naked or with 5 small projections. Akenes mostly stony.

            Lobes of the tubular corolla acute and erect, (*Onosmòdium*) FALSE-GROMWELL.

            Lobes of the trumpet-shaped corolla spreading, rounded, short. Akenes fleshy. Plant very smooth, (*Merténsia*) LUNGWORT.

Lobes of the salver-shaped or funnel-shaped corolla spreading, rounded.
   Each with one edge outside and one inside in the bud : corolla very
      short, ·   (*Myosótis*) SCORPION-GRASS or FORGET-ME-NOT.
   Two lobes covering the others in the bud.
      Corolla short, white or whitish, funnel-shaped,   (*Lithospérmum*) GROMWELL.
      Corolla long, orange-yellow, salver-shaped, (*Lithospérmum*, § *Bátschia*) PUCCOON.
Akenes or lobes of the ovary prickly, fixed by their side or upper end to the base of the
               style. Corolla salver-shaped, with 5 scales in the throat.
   Erect, prickly on the margins only. Flowers small,   (*Echinospérmum*) STICKSEED.
   Oblique or flattened from above, short-prickly or rough all over, (*Cynoglóssum*) HOUND'S-TONGUE.

Ovary not lobed, but splitting when ripe into 4 akenes: corolla short, (*Heliotròpium*) *HELIOTROPE.

65. WATERLEAF FAMILY. Order HYDROPHYLLACEÆ.

Herbs with lobed, compound, or toothed and mostly alternate leaves ; the regular flowers
much like those of the Borrage Family, except as to the ovary, which is globular and only
one-celled and bears the
few or many ovules and
seeds on the walls (pari-
etal), or on two projections
from them. In Waterleaf,
Nemophila, &c., the two
placentas, bearing the few
seeds, broaden and make
a kind of lining to the
pod. Corolla bell-shaped

 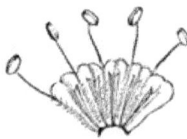 

452. Flower of Virginia Waterleaf.   453. Corolla laid open, and stamens.
       454. Calyx and young pod, with the style.

or wheel-shaped ; its lobes and the stamens always 5.   Style 2-cleft above.   The Waterleaf furnishes our principal plants of the family that are common wild. But some Ne-
mophilas and Phacelias, from Texas and California, are showy garden annuals.

Leaves opposite, at least the lower ones. Stamens not projecting beyond the corolla.
   Calyx without appendages or teeth between the divisions, large in fruit,   (*Ellisia*) ELLISIA.
   Calyx with 5 reflexed teeth between the divisions,   (*Nemóphila*) *NEMOPHILA.
Leaves alternate: appendages of the calyx none or minute: stamens long.
   Mostly annuals: seeds on the walls of the pod, or two narrow placentas,   (*Phacèlia*) PHACELIA.
   Perennials, with scaly-toothed rootstocks.   Seeds 1 to 4, enclosed in a membrane
            which lines the pod.   Flowers white or bluish, clustered: filaments
            bearded below,   (*Hydrophýllum*) WATERLEAF.

Waterleaf. *Hydrophýllum.*

1. VIRGINIA W. Smoothish, 1° or 2° high; leaves pinnately divided into 5 or 7 narrow and toothed
   or cleft lobes; calyx hairy. Rich woods.   *H. Virginicum.*
2. CANADA W. Smoothish ; leaves rounded, palmately lobed, longer than the peduncle ; calyx
   smooth. Rich woods.   *H. Canadense.*

66. POLEMONIUM FAMILY. Order POLEMONIACEÆ.

Herbs, not twining (but Cobæa climbs by tendrils), with regular flowers, all the parts in fives, except the pistil, which is 3-celled and the style 3-cleft at the top, the 5 spreading lobes of the corolla *convolute* in the bud, i. e. overlapping so that one edge of each is outside of that behind it, but inside of the next one. Flowers generally handsome. All the kinds here given are cultivated; but the Phloxes are wild in this country (especially W. and S.), and so is one Polemonium. Gilias are pretty garden annuals from California, &c. Cobæa, which is placed here, though very different from the rest, is a great-flowered vine from Mexico.

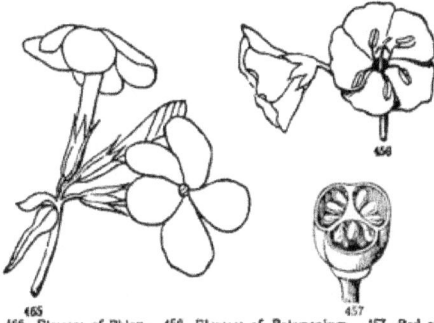

455. Flowers of Phlox. 456. Flowers of Polemonium. 457. Pod of Polemonium, cut across.

Climbing by tendrils on the pinnate leaves: flowers axillary, single: calyx leafy: corolla
    bell-shaped, large, but dull-colored,   (*Cobæa*) *COBÆA.
Not climbing: flowers in panicled cymes or clusters.
    Stamens inserted at very unequal heights on the long tube of the salver-shaped
    corolla, short, included: calyx narrow, 5-angled: seeds only one in each cell.
        Leaves all entire, sessile, and opposite, except the uppermost,   (*Phlox*) PHLOX.
    Stamens all inserted at the same height. Leaves mostly alternate and compound.
        Corolla almost wheel-shaped (light blue): stamens turned towards the lower side
            of the flower: leaves pinnate,   (*Polemònium*) POLEMONIUM.
        Corolla funnel-shaped or salver-shaped: stamens not turned to one side: seeds
        several. Leaves once to thrice pinnately divided,   (*Gilia*) GILIA.

Phlox. *Phlox.*

    * Perennial herbs, growing in open woods, and in gardens.

1. PANICLED P. Stem stout, 2° to 4° high; leaves lance-oblong and ovate-lanceolate, pointed, tapering or the upper ones heart-shaped at the base; panicle large and broad; corolla pink or white, the lobes entire. Fl. summer.   *P. paniculàta.*

2. SPOTTED P. Stem 1° or 2° high, slender, simple, purple-spotted; lower leaves lance-shaped, uppermost lance-ovate, tapering upwards from the rounded or slightly heart-shaped base; panicle narrow; calyx-teeth rather blunt; corolla pink-purple, or varying to white in gardens, the lobes entire. Fl. summer.   *P. maculàta.*

3. HAIRY P. Stems slender, ascending, 1° or 2° high, clammy-hairy; leaves lance-shaped or lance-linear; cyme flat; calyx-teeth long, awn-pointed; lobes of the rose-pink corolla entire. Fl. early summer.   *P. pilòsa.*

4. RUNNING P. Spreading by creeping runners, bearing roundish and thickish smooth leaves; flowering stems 4' to 8' high, with oblong leaves; flowers few and large; lobes of the red-purple corolla round and entire. Fl. early summer.   *P. réptans.*

5. SPREADING P. Stems ascending, 9' to 18' high, rather clammy; leaves ovate-oblong or broad lance-shaped; cyme loosely-flowered; lobes of the pale lilac or bluish corolla generally obcordate and rather distant from each other. Fl. spring, N. & W.   *P. divaricàta.*

6. GROUND P. or MOSS-PINK. Plant creeping and tufted in flat mats; leaves awl-shaped or lance-linear, small, crowded; corolla pink or rose-color, with a darker eye, sometimes white. Fl. spring, in sandy or rocky soil. S. & E.   *P. subulàta.*

* * Garden annual from Texas.

7. DRUMMOND'S P. Rather clammy, branched; leaves lance-oblong, the upper heart-shaped at the base; corolla crimson, purple or rose-color, lobes entire.   *P. Drummóndii.*

Polemonium. *Polemònium.*

1. BLUE P. (Called in gardens *Jacob's Ladder* or *Greek Valerian.*) Stem erect, 1° or 2° high, leafy; leaflets many; seeds several. Gardens.   *P. cærùleum.*

2. WILD P. Stems weak, spreading; leaflets 7 to 11; flowers few. Woods, W. & S.   *P. reptans*

67. CONVOLVULUS FAMILY. Order CONVOLVULACEÆ.

Twining or trailing herbs, often with some milky juice, with alternate leaves and regular flowers: calyx of 5 sepals: corolla 5-plaited or 5-lobed. Stamens 5. Pistil making a round pod, with 2 to 4 cells and one or two large seeds erect from the bottom of each cell. (For illustrations see Fig. 4 to 7, 13 to 22.) Dodders are leafless parasitic plants of the family.

Plants with foliage, and bearing large flowers, open only for one day. Style one.
   Stamens protruded beyond the mouth of the tubular or trumpet-shaped and crimson
        or scarlet corolla,   (*Quámoclit*) QUAMOCLIT.
   Stamens included in the tube of the almost entire corolla.
      Stigma thick, 2-lobed: corolla bell-shaped: pod 4-celled, 4-seeded, (*Batàtas*) SWEET-POTATO.
      Stigma capitate, thick, with 2 or 3 lobes: corolla funnel-form: pod with 2 or 3
        cells, and 2 seeds in each cell,   (*Ipomæa*) MORNING-GLORY.
      Stigmas 2, long, linear or oblong.   [BINDWEED.*
        Calyx naked at the base: corolla bell-shaped,   (*Convólvulus*) *CONVOLVULUS or
        Calyx covered by 2 large bractlets: corolla funnel-form, (*Calystègia*) BRACTED-BINDWEED.
Plants with leafless whitish, reddish, or yellowish thread-like stems, twining over other
   plants, and attaching themselves to their bark, on which they feed: flowers in clusters: corolla bell-shaped, with 5 scales inside the stamens: pod 2-celled, cells 2-seeded:
   embryo spiral, without any cotyledons,   (*Cùscuta*) DODDER.

Quamoclit. *Quámoclit.*

1. CYPRESS-VINE Q. Leaves narrow, pinnately dissected into thread-shaped divisions; limb of the corolla rather deeply 5-lobed. Garden annual.   *Q. vulgàris.*

2. SCARLET Q. Leaves heart-shaped, entire or nearly so; corolla scarcely lobed,   *Q coccínea.*

---

* The low THREE-COLORED CONVOLVULUS (*C. tricolor*) is a garden annual.

Morning-Glory. *Ipomǽa.*

1. COMMON M. Annual; stem hairy, the hairs bent downwards; leaves heart-shaped, entire; flowers 3 to 5 on the peduncle; flowers purple or pink varying to white, opening early in the morning, closing in bright sunshine; pod 3-celled. Cult. &c. *I. purpùrea.*

2. WILD M. (or MAN-OF-THE-EARTH). Smooth; root huge, perennial; leaves heart-shaped, entire or some of them narrowed in the middle; flowers 1 to 5 on a peduncle, white with purple in the tube, opening in sunshine. Sandy banks. *I. pandurátus.*

68. NIGHTSHADE FAMILY. Order SOLANACEÆ.

Herbs, or sometimes shrubs, with a colorless bitter or nauseous juice (often poisonous); alternate leaves; and regular flowers, with 5 (or in cultivated plants sometimes 6 or 7) mostly equal stamens and one pistil. Ovary with 2 or more cells, in fruit becoming a many-seeded berry or pod. Corolla plaited in the bud, or *valvate*, i. e. the lobes placed edge to edge.

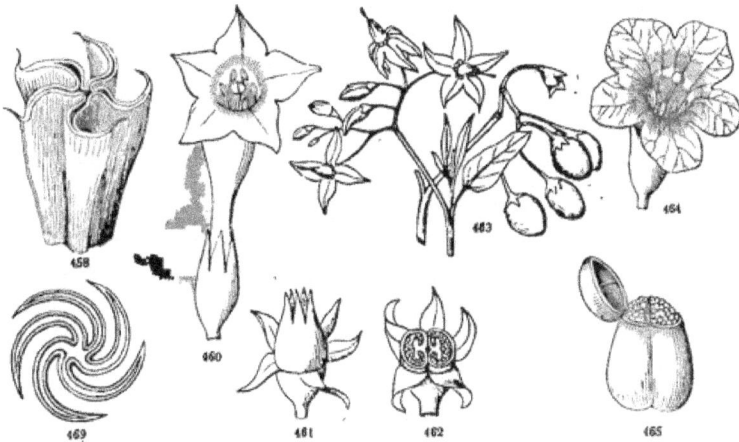

458. Upper part of the corolla of Stramonium (Fig. 177) in bud 459. Cross-section of the same, to show how it is plaited and folded. 46). Flower of Tobacco. 461. Its pod and calyx. 462. Same, with the upper part cut away 463. Flowers and berries of Bittersweet Nightshade. 464. Flower of Henbane. 465. Pod of the same, opening by a lid.

Corolla wheel-shaped : stamens closely converging or united around the style (Fig. 182, 183). Fruit a berry.

    Anthers longer than the very short filaments, and

        Connected with each other, opening lengthwise. Berry several-celled, (*Lycopérsicum*) *TOMATO.

        Not grown together, opening at the top by two pores,   (*Solànum*) NIGHTSHADE.

    Anthers shorter than the filaments, heart-shaped, separate, opening lengthwise. Berry

        pod-like, inflated, the pulp very pungent (Cayenne or Red Pepper),

                                    (*Cápsicum*) *CAPSICUM.

Corolla between wheel-shaped and bell-shaped, or very open and short funnel-shaped,
with an almost entire border: anthers separate, shorter than the filaments: calyx enlarged and enclosing the berry.

Calyx 5-lobed, becoming a bladdery bag around the (eatable) berry, (*Physalis*) GROUND-CHERRY.

Calyx 5-parted, the divisions becoming heart-shaped: berry dry, (*Nicándra*) *APPLE-OF-PERU.

Corolla funnel-shaped, bell-shaped, or tubular: stamens separate: filaments slender.

Calyx 5-parted, leafy, spreading: stamens curved or unequal.

Corolla bell-shaped : stamens curved: fruit a black berry (deadly poisonous), (*Átropa*) *DEADLY NIGHTSHADE.

Corolla funnel-shaped: stamens unequal: fruit a pod, (*Petùnia*) *PETUNIA.

Calyx 5-toothed or 5-lobed.

Shrubby, with vine-like branches and narrow leaves : corolla funnel-shaped,
small: fruit a berry, (*Lýcium*) *MATRIMONY-VINE.

Herbs (annuals), unpleasant-scented, mostly large-flowered. Fruit a pod.

Corolla (dull and veiny) and stamens rather irregular: pod in the urn-shaped
calyx, opening at the top by a lid (Fig. 465), (*Hyoscýamus*) HENBANE.

Corolla perfectly regular, generally long funnel-shaped.

Calyx 5-angled, long, falling away after flowering : pod large and
prickly, 2-celled and becoming 4-celled, 4-valved. (Flower,
Fig. 177, 458), (*Datùra*) STRAMONIUM.

Calyx not angled, remaining around the smooth pod, which opens by
several slits at the top, (*Nicotiàna*) *TOBACCO.

The only genus which needs to have the species enumerated is the

Nightshade. *Solànum.*

* Anthers blunt: plants not prickly.

1. COMMON NIGHTSHADE. A very common low, much-branched, homely weed, in damp or shady
grounds ; root annual; leaves ovate, wavy-toothed; flowers very small, white; berries black,
small, said to be poisonous. *S. nigrum.*

2. BITTERSWEET N. Stem rather shrubby, climbing; leaves ovate and heart-shaped, some of them
halberd-shaped or with an ear-like lobe at the base on one or both sides; flowers blue-purple,
in small cymes; berries bright red. Around dwellings, &c. (The flowers are represented in Fig.
182, as well as Fig. 463.) *S. Dulcamàra.*

3. JERUSALEM-CHERRY N. A low tree-shaped shrub, with lance-oblong and smooth entire leaves,
scattered and small white flowers, succeeded by large bright red berries like cherries. Cultivated
in houses, &c. *S. Pseudo-Cápsicum.*

4. POTATO or TUBEROUS N. Shoots under ground bearing tubers (Fig. 60); leaves interruptedly
pinnate; the leaflets very unequal, some of them minute; corolla only 5-angled (Fig. 183), white
or blue. Cultivated. *S. tuberòsum.*

* * Anthers long and taper-pointed: stems and leaves prickly.

5. EGG-PLANT N. Leaves ovate, wavy or somewhat lobed, downy; berry oblong, purple or whitish,
from the size of an egg to that of a melon, eatable when cooked. Cult. *S. Melongèna.*

6. HORSE-NETTLE N. Leaves ovate or oblong, wavy or angled, hoary-hairy; corolla bluish; berry
yellow. A weed, S. *S. Carolinénse.*

69. GENTIAN FAMILY. Order GENTIANACEÆ.

Smooth herbs with a colorless bitter juice; the leaves, with two exceptions, opposite, sessile, and entire; the regular flowers having as many stamens as there are lobes to the corolla, and alternate with them; stigmas or branches of the style 2; pod one-celled, with many and usually very small seeds on the walls, usually in two lines. — Tonic, generally very bitter plants: none of them poisonous. Flowers commonly large and handsome.

Leaves simple, opposite and sessile. Corolla with the lobes *convolute*, i. e. each with one
      edge in and one out, in the bud.
    Corolla wheel-shaped, 5- to 12-parted, white or pink, in cymes. Style 2-parted.
      (Two or three handsome-flowered species in salt marshes, and one or two
      on river-banks, &c., especially South),   (*Sabbátia*) SABBATIA.
    Corolla funnel-form or bell-shaped, commonly blue. Style very short or none: stig-
      mas 2, broad,   (*Gentiàna*) GENTIAN.
Leaves simple, alternate or all from the root, round-heart-shaped, floating on the water,
      with very long footstalks, which bear near their summit a cluster of small
      white flowers, along with some spur-shaped bodies. Corolla 5-parted, the
      lobes folded inwards in the bud,   (*Limnánthemum*) FLOATING-HEART.
Leaves with 3 oblong leaflets; footstalks long, alternate, their base sheathing the thickish
      rootstock or the lower part of a scape, which bears a raceme of white
      flowers. Corolla 5-parted, the lobes white-bearded inside, their edges
      turned inwards in the bud. One species, in bogs,   (*Menyánthes*) BUCKBEAN.

Gentian. *Gentiàna.*

   * Stamens separate: no plaits or fringes between the lobes of the corolla.

1. FIVE-FLOWERED GENTIAN. Slender, branching; leaves lance-ovate; branches about 5-flowered; corolla light blue, hardly 1′ long, with 5 pointed naked lobes. Fl. late summer and autumn; as do all the species.   *G. quinqueflóra.*

2. FRINGED G. Leaves lance-shaped or lance-ovate; flowers single on a long naked stalk; corolla 2′ long, sky-blue, with 4 obovate beautifully fringed lobes. Low grounds.   *G. crinita.*

   * * Anthers cohering with each other more or less: corolla with 5 plaited folds.

3. CLOSED G. Stout, leafy to the top, the flowers in sessile clusters, terminal and in the axils of the upper lance-oblong leaves; corolla pale blue or purplish, rather club-shaped, with the mouth contracted, and with 5 fringe-toothed plaits, the lobes hardly any.   *G. Andréwsii.*

4. SOAPWORT G. The light blue corolla more open and bell-shaped, its lobes short and broad, but longer than the intervening plaits; otherwise much as No. 3. S. and W.   *G. Saponària.*

5. WHITISH G. Leaves lance-ovate with a heart-shaped clasping base; corolla dull white or yellowish, with lobes longer than the plaits. S. and W.   *G. alba.*

70. DOGBANE FAMILY. Order APOCYNACEÆ.

Plants with a milky and acrid juice, a tough inner bark, generally opposite and entire leaves, and regular flowers: corolla 5-lobed, the lobes *convolute* in the bud (one edge in,

13

the other out) ; the 5 stamens on the corolla alternate with its lobes; the anthers generally more or less adherent to the stigma. Ovaries 2 ; but the stigmas, and often the styles also, united into one ; the fruit two separate pods. Seeds generally many, and with a tuft of down at one end.

Corolla with a funnel-shaped tube and a wheel-shaped 5-parted border: style one.

     Leaves generally in whorls. Shrub, with large rose-colored flowers,   (*Nérium*) *OLEANDER.
     Leaves opposite, evergreen in the common creeping species. Fl. blue,   (*Vínca*) *PERIWINKLE.
     Leaves alternate, very many, narrow. Erect herbs with pale-blue salver-shaped flow.
         ers: seeds not tufted,   (*Amsónia*) AMSONIA.
Corolla bell-shaped, white or pinkish: style none. Herbs, with opposite leaves. (*Apócynum*) DOGBANE.

Dogbane. *Apócynum.*

1. SPREADING D. Branches of the low erect stem widely diverging; leaves ovate or oval; cymes few-
flowered; lobes of corol-
la recurved; tube shorter
than the calyx. Thickets,
&c. *A. androsœmifólium.*

2. HEMP D. or INDIAN HEMP.
Stem and branches erect
or ascending; cymes few-
flowered; lobes of the corolla not recurved, the
tube not longer than the
calyx. *A. cannábinum.*

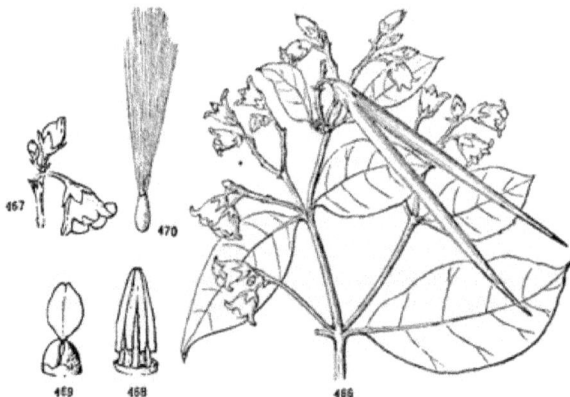

466. Summit of a plant of Dogbane,
No. 1, with flowers and pods
467. Flowers, enlarged.
468. Flower with the corolla cut
away, to show the stamens.
469. The stamens taken away, to show
the pistils ; two ovaries, with their two
large stigmas united into one mass.
470. A seed, with its tuft of long hairs
or down at one end.

71. MILKWEED FAMILY.   Order ASCLEPIADACEÆ.

Plants with milky juice, tough bark, and in other respects like the Dogbane family, but with the 5 short stamens all united by their filaments into a ring or tube, the anthers grown fast to the large stigma, and the grains of pollen in each cell cohering into a waxy or tough mass. Flowers in simple umbels. Pods a pair of many-seeded follicles : seeds furnished with a long tuft of silky down at one end (Fig. 229). The flowers in this family are curious, but are too difficult for the beginner. The two common genera may be distinguished as follows : —

Corolla 5-parted, reflexed: five hoods to the stamens, with a horn in each,   (*Asclépias*) MILKWEED.
Corolla, &c. as in Milkweed, but the hoods without any horn,   (*Acerátes*) GREEN-MILKWEED.

72. JESSAMINE FAMILY. Order JASMINACEÆ.

Shrubby, mostly climbing plants, with opposite and mostly compound (pinnate) leaves, and perfect flowers with a salver-shaped corolla of 5 or more lobes overlapping in the bud, but only 2 stamens. Ovary 2-celled, with 2 or 3 ovules erect from the base of each cell. No wild species; but in gardens and houses we have the common (*Jásminum*) *JESSAMINE.

73. OLIVE FAMILY. Order OLEACEÆ.

Shrubs or trees, with opposite leaves; the corolla, when there is any, 4-lobed, and the lobes *valvate* (edge to edge) in the bud, but the stamens only 2 and short: sometimes there are 4 distinct petals; and all our species of Ash are without petals. Ovary 2-celled, with 2 ovules hanging from the top of each cell: the fruit often one-celled and one-seeded; either a stone-fruit, as in the Olive and Fringe-tree; a berry, as in Privet; a pod, as in Lilac; or a key, as in the Ash.

Corolla salver-shaped or funnel-shaped, with a 4-lobed border: flowers perfect, in thick
    panicles. Leaves simple, entire.
    Corolla salver-shaped with a long tube: fruit a flat 4-seeded pod,   (*Syringa*) *LILAC.
    Corolla short, funnel-shaped: fruit a 1- or 2-seeded berry. Low shrub,   (*Ligústrum*) *PRIVET.
Corolla of 4 very long and narrow petals, barely united at the bottom. Drupe one-seeded.
    Low tree or shrub, with simple leaves, and slender drooping panicles of delicate
    snow-white blossoms,   (*Chionánthus*) FRINGE-TREE.
Corolla none: even the calyx small or sometimes none: stamens 2, rarely 3 or 4, on the
    receptacle: fruit a key, winged at the top or all round, one-seeded. Trees, with
    opposite pinnate leaves,   (*Fráxinus*) ASH.

Lilac. *Syringa.*

1. COMMON LILAC. Leaves more or less heart-shaped; flowers lilac or white, in spring. Cultivated: one of the commonest ornamental shrubs.   *S. vulgàris.*
2. PERSIAN LILAC. Leaves oblong or lance-shaped; clusters more slender. Cultivated.   *S. Pérsica.*

Ash. *Fráxinus.*

The flowers in all our species appear in early spring, in clusters, and are diœcious, or nearly so.
    *.Key winged from the top only: leaflets stalked.

1. WHITE ASH. Shoots and stalks smooth; leaflets 7 to 9, pale (smooth or downy) beneath; body of the key marginless and blunt.   *F. Americàna.*
2. RED ASH. Shoots and stalks velvety; leaflets 7 to 9, downy beneath; body of the key 2-edged, acute at the base, the wing long and narrow.   *F. pubéscens.*
3. GREEN ASH. Smooth throughout; leaflets 5 to 9, green both sides; key as in No. 2.   *F. víridis.*
    * * Key winged all round, oblong.
4. BLACK ASH. Leaflets 7 to 11, sessile; oblong-lanceolate, tapering to a point, green both sides; no calyx to the fertile flowers. Swamps; common N.   *F. sambucifòlia.*
5. BLUE ASH. Branchlets square; leaflets 7 to 9, short-stalked, lance-ovate. W.   *F. quadrangulàta.*

III. Apetalous Division.

74. BIRTHWORT FAMILY. Order ARISTOLOCHIACEÆ.

Herbs or twining vines, with perfect and large flowers, the tube of the 3-lobed calyx
coherent with the 6-celled
and many-seeded ovary.
Leaves   mostly   heart-
shaped or kidney-shaped,
and entire, on long foot-
stalks, alternate, or else
from the rootstock at the
surface of the ground.
Lobes of the calyx edge
to edge in the bud, usu-
ally dull-colored.

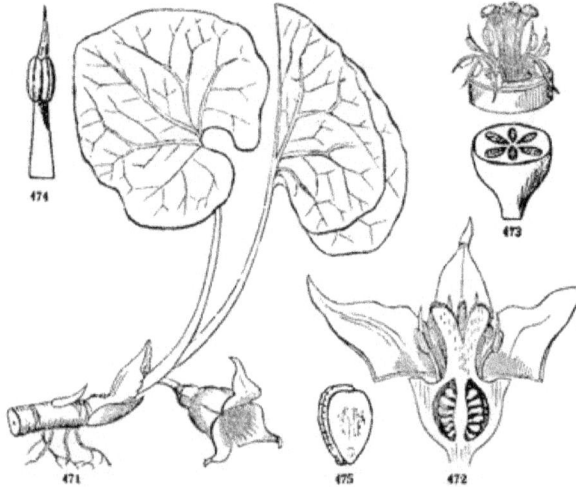

471. Plant of Canada Asarum or
Wild-Ginger, in flower. 472. Magnified flower divided lengthwise, and the
calyx spread out flat. 473. Flower,
with the lobes of the calyx cut away,
and the ovary cut across. 474 A separate stamen, more magnified; outside
view. 475. Magnified seed divided
lengthwise.

Stemless herbs, with a pair of leaves and a flower between them from the spicy-tasted
    and creeping rootstock: calyx short, 3-cleft or 3-lobed; stamens 12, with filaments,
    which are united only with the base of the thick 6-lobed style, and are pointed above
    the anthers,   (*Asarum*) WILD-GINGER.
Twining shrubs or else low herbs: calyx a crooked tube, with a narrow throat and a
    slightly 3-lobed spreading border: stamens 6, sessile on the outside of the 3 lobes of
    the sessile stigma, i. e. two anthers or 4 cells to each lobe, attached to the stigma
    by their whole length: fruit a 6-valved pod, filled with numerous flat seeds,
                                                          (*Aristolochia*) BIRTHWORT.

Birthwort. *Aristolòchia.*

1. SNAKEROOT B. or VIRGINIA SNAKEROOT. Herb 8' to 15' high; several stems from a tufted root,
   downy; flowers borne next the ground, in general shape much like the letter S; leaves oblong-
   heart-shaped or halberd-shaped. Rich woods; becoming scarce.   *A. serpentària.*
2. PIPE-VINE B. A tall woody climber, with rounded kidney-shaped leaves, 8' or 12' broad when
   full grown; flower 1½' long, curved like a Dutch pipe ; greenish outside, and with the short
   3-lobed border brown-purple within. Alleghany Mountains, or near them; and cultivated for
   arbors.   *A. Sipho.*

75. MIRABILIS FAMILY. Order NYCTAGINACEÆ.

Has some wild representatives far west and south, viz.: OXYBAPHUS, &c., with several flowers in a calyx-like involucre, the funnel-shaped calyx rose-purple, and exactly like a corolla. And in gardens MIRABILIS or FOUR-O'CLOCK (so called from the flowers opening late in the afternoon) is common. Here there is only one flower in the bell-shaped involucre, which exactly imitates a calyx, while the large funnel-shaped calyx is just like the corolla of a Morning-Glory. Stamens 5 : style one. Leaves opposite, heart-shaped, long-stalked. The

COMMON FOUR-O'CLOCK or MIRABILIS, from Mexico, well known in gardens, is *M. Jalápa.*

76. POKEWEED FAMILY. Order PHYTOLACCACEÆ.

Is represented with us by one, and that a very common, species of

Pokeweed. *Phytolácca.*

Sepals 5, rounded, concave, petal-like, white. Stamens 10, under the ovary. Ovary green, composed of 10 one-seeded ovaries united into one: styles 10, short and separate. Fruit a dark crimson 10-seeded berry. A coarse rank herb, with a thick, acrid, and poisonous root, a large pithy stem, and alternate oblong leaves ; the flowers in racemes opposite the leaves. Low and rich ground, everywhere common; flowering all summer, ripening its abundant berries in autumn.
P. decándra.

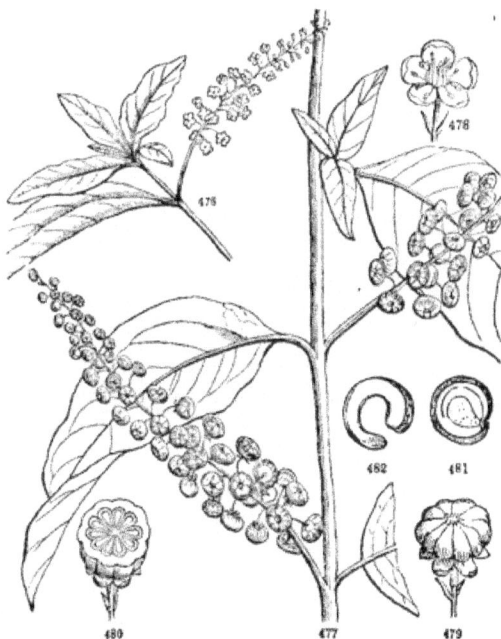

476. Summit of a flowering branch of Poke-
    weed.
477. Fruit-bearing branch.
478. A flower, enlarged.
479. Young fruit.
480. Same, cut across.
481. Seed divided lengthwise, and magnified.
482. Embryo, more magnified.

77. GOOSEFOOT FAMILY. Order CHENOPODIACEÆ.

Homely herbs, with mostly alternate leaves, without stipules, and no dry scaly bracts among the small and greenish flowers ; the calyx enclosing the one-celled and one-seeded ovary, but not adhering to it, and bearing from one to five stamens. Styles 2 to 5, short. Weeds (several called PIGWEEDS), abounding in cultivated or waste grounds, and some are pot-herbs. The small flowers and fruits make them too difficult for the beginner. The following key will lead the student to the name of the principal common kinds.

Leafless fleshy herbs, in salt marshes, with perfect flowers in fleshy spikes,   (*Salicórnia*) SAMPHIRE.
Leafy herbs, with broad or broadish, generally tender leaves, not prickly: calyx wingless.
    Flowers perfect.
       In clusters or spiked heads: calyx becoming berry-like, altogether making a
             strawberry-like red pulpy fruit,   (*Blitum*) BLITE.
       In small sessile clusters collected in spikes or panicles: calyx dry and herba-
            ceous.
           Akene thick and hard, below adherent to the calyx. Leaves smooth,   (*Beta*) *BEET.
          Akene very thin and breaking away from the seed. Leaves often mealy.
             Pigweeds,   (*Chenopódium*) GOOSEFOOT.
    Flowers monœcious: the fertile ones single in the axils of the leaves. Sea-coast, and
       one rarely cultivated as a pot-herb,   (*A'triplex*) ORACHE.
    Flowers diœcious, in spiked clusters: calyx over the fruit, with 2 to 4 horns or pro-
       jections: leaves arrow-shaped,   (*Spinàcia*) *SPINACH.
Leafy and much-branched plants on the sea-shore; the leaves awl-shaped and prickly-
       tipped: flowers perfect: calyx winged in fruit,   (*Salsòla*) SALTWORT.

78. AMARANTH FAMILY.   Order AMARANTACEÆ.

Herbs, much like the last family in almost every character, except that the flowers are furnished with 3 or more dry and scale-like thin bracts: these are sometimes brightly colored, so as to make showy clusters or bunches, and, being dry, they do not wither after blossoming. The little one-seeded pod in many cases is a pyxis (242), that is, it opens round the middle, the upper part falling off, as a lid. The common species belong mainly to two genera: —

183. Pod of Am-
aranth opening by
a lid.

Flowers in spiked or panicled clusters, terminal or axillary: stamens 5 or 3, separate:
    little pod opening by a lid. To this belongs one kind of PIGWEED, and the
    PRINCE'S FEATHER, LOVE-LIES-BLEEDING, COXCOMB, &c., in gardens and
    enriched soil,   (*Amarántus*) AMARANTH.
Flowers in a head: stamens 5, monadelphous, and the filaments 3-cleft, the middle lobe
    bearing the anther,   (*Gomphrèna*) *GLOBE-AMARANTH.

79. BUCKWHEAT FAMILY.   Order POLYGONACEÆ.

Herbs with alternate entire leaves, and mostly perfect flowers; with a calyx of 4 to 6 sepals (separate or united at the base), and 3 to 9 stamens inserted on its base: ovary one-celled making a one-seeded akene; its styles or stigmas 2 or 3. Besides, this family may always be known by the stipules which form a sheath above each joint (as in Fig. 137). The watery juice is often sour, as in Rhubarb and Sorrel, sometimes sharp and biting.

Calyx of 5 (rarely 4) nearly similar sepals, all more or less petal-like.

Stamens 4 to 9: akene generally small: cotyledons narrow,   (*Polygonum*) KNOTWEED.

Stamens 8: styles 3: akene triangular, shaped like a beechnut, much longer than the calyx: cotyledons very broad and folded in the mealy albumen: root annual: leaves nearly halberd-shaped: flowers white, corymbed,   (*Fagopyrum*) *BUCKWHEAT.

Calyx of 6 sepals, and

All alike and petal-like (white): stamens 9: styles 3,   (*Rheum*) *RHUBARB.

Three outer ones herbaceous and spreading: three inner larger, especially after flowering, when they close over the triangular akene: flowers diœcious: leaves sour, eared or halberd-shaped,   (*Rumex, § Acetosella*) SORREL.

Flowers perfect or polygamous: leaves bitter: coarse herbs,   (*Rumex*) DOCK.

Knotweed. *Polygonum.*

* Flowers single or several together in the axils of the leaves, greenish or whitish: sheaths (stipules) cut-fringed or torn into narrow shreds.

1. COMMON KNOTWEED, KNOTGRASS, or GOOSEGRASS. Spreading on the ground, small; leaves sessile, lance-shaped or oblong, pale; a variety has nearly upright stems and oblong or oval leaves. The commonest weed in yards and waste places.   *P. aviculàre.*

2. SLENDER K. Upright, somewhat branched; leaves linear, acute, sheaths fringed. Dry soil. *P. ténue.*

* * Flowers in terminal heads, spikes, or racemes.

← Not twining nor climbing, and leaves not heart-shaped nor arrow-shaped: calyx petal-like and 5-parted, except in No. 10.

3. ORIENTAL K. or PRINCE'S FEATHER. Tall annual, 4° to 7° high; leaves ovate; spikes of rose-colored flowers long and nodding; stamens 7; akene flattish. Gardens.   *P. orientàle.*

4. WATER K. Stems floating in water, or rooting in mud, or upright; leaves lance-shaped or oblong; spike thick and short; flowers rose-red; stamens 5; styles 2.   *P. amphíbium.*

5. PENNSYLVANIA K. Stem upright, 1° to 3° high; leaves lance-shaped; spike oblong, thick, erect, its peduncle beset with club-shaped bristles or glands; flowers rose-colored; stamens 8; akene flat. Moist ground.   *P. Pennsylvánicum.*

6. LADY'S-THUMB K. Stems, &c. like the last and next, but no bristles on the peduncle; leaves with a darker spot on the upper side; spike short and thick, erect; flowers greenish-purple; stamens 6. Very common in waste places.   *P. Persicària.*

7. SMARTWEED or WATER-PEPPER K. Upright, annual, 1° or 2° high, very acrid and biting to the taste; leaves and also the greenish sepals marked with fine transparent dots; spikes short but loose, drooping; akene flattish or bluntly triangular. Moist ground, common in waste places, yards, and near dwellings.   *P. Hydropiper.*

8. WILD SMARTWEED K. Upright, 1° to 3° high from a perennial root, biting like the last, and the leaves dotted; spikes very slender, erect, whitish or flesh-color; stamens 8; styles 3; akene sharply triangular. Wet places.   *P. acre.*

9. MILD WATER-PEPPER K. Upright, 1° to 3° high; often creeping at the base and rooting in water; leaves roughish, not biting, narrowly lance-shaped; spikes slender, erect, rose-color; stamens 8; style 3-cleft at the top; akene sharply triangular. Shallow water.   *P. hydropiperoides.*

10. VIRGINIA K. Stem 2° to 4° high, angled; leaves large, ovate or lance-ovate, taper-pointed; flow-

ers scattered in a long and naked slender spike; calyx greenish, 4-parted; stamens 5; styles 2, bent down in fruit. Thickets. *P. Virginiànum.*

＋ ＋ Somewhat climbing, or supported by recurved sharp prickly bristles on the strong angles of the stems, &c.; flowers white or flesh-color in small racemes or heads; root annual. The prickly angles cut like a saw, whence the plants aro called *Tear-Thumb.*

11. ARROW-LEAVED K. Leaves arrow-shaped (Fig. 100), short-stalked; akene 3-angled. *P. sagittàtum.*

12. HALBERD-LEAVED K. Leaves halberd-shaped (Fig. 102), long-stalked ; akene flattish. Low grounds. *P. arifòlium.*

＋ ＋ ＋ Twining annuals, with smooth stems and greenish or whitish flowers in panicled racemes; leaves heart-shaped and partly halberd-shaped.

13. CLIMBING K. Smooth, climbing high over shrubs, &c.; racemes leafy; 3 of the calyx-lobes more or less winged in fruit. Thickets in low ground. *P. dumetòrum.*

14. BINDWEED K. Low, stems roughish; racemes corymbed; three of the calyx-lobes ridged in the middle. Cult. and waste grounds. *P. Convòlvulus.*

80. LAUREL FAMILY. Order LAURACEÆ.

Trees or shrubs, with spicy bark and leaves; the latter marked with transparent dots under a magnifying-glass, alternate and simple; the calyx of 6 petal-like sepals. Stamens 9 or 12 on the very bottom of the calyx ; the anthers opening by up-lifted valves. Pistil simple, with a one-celled ovary, in fruit forming a berry or drupe, one-seeded. Flowers generally polygamous or diœcious in spring. — A very well-marked family, mostly in hot countries, but we possess two or three representatives.

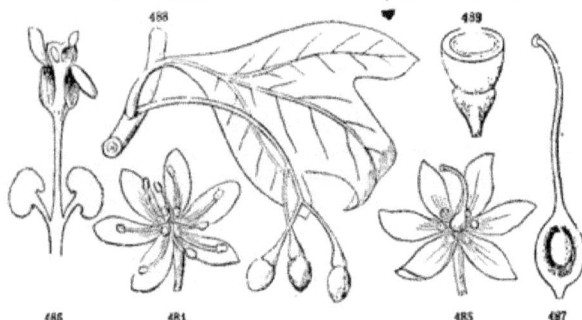

484. Sterile flower of Sassafras 485. Fertile flower of the same. 486 Magnified stamen, with two glands at the base , the anther opening by two large and two small valves 487 Pistil, with the ovary divided to show the ovule hanging from the top. 488. Leaf and cluster of fruit. 489. Lower half of fruit, cut across.

Flowers perfect : stamens 9, with good anthers, and 3 sterile ones. Tree, with entire oblong leaves; common South, (*Pèrsea*) RED-BAY.

Flowers diœcious or nearly so, greenish-yellow: stamens 9, about 3 of them with yellow glands at the base of the filaments (Fig. 486).

Anthers 4-celled and 4-valved. Tree: flowers in stalked corymbs, appearing with the leaves: some of the latter 3-lobed, (*Sàssafras*) SASSAFRAS.

Anthers 2-celled and opening by a single valve to each cell. Shrub: flowers in sessile clusters, appearing earlier than the entire leaves, (*Bénzoin*) SPICEBUSH.

81. MEZEREUM FAMILY. Order THYMELEACEÆ.

Shrubs, with very tough and acrid bark; entire generally alternate leaves; and perfect flowers, with a tubular calyx colored like a corolla, bearing 8 or 10 stamens, free from the simple pistil. Ovary one-celled, one-ovuled, making a berry in fruit.—We have one wild plant of the family; *Daphne Mezereum* is a hardy low shrub in gardens, and *D. odora* in houses. Flowers appearing earlier than the leaves.

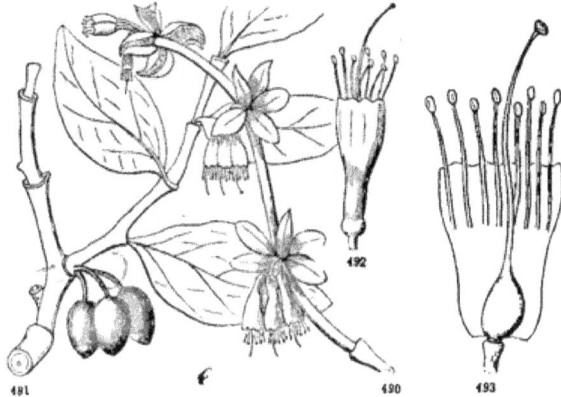

490. Flowering branchlet of Leatherwood. 491. Branch with foliage and fruit. 492. A flower, magnified. 493. Same, more magnified, the calyx laid open.

Calyx salver-shaped or funnel-shaped, generally rose-color, the border 4-lobed: stamens 8, in two sets, included; filaments hardly any, (*Daphne*) *DAPHNE.

Calyx tubular, pale yellow, with no spreading border, obscurely 4-toothed: stamens 8, with long protruded filaments, (*Dirca*) LEATHERWOOD.

82. NETTLE FAMILY. Order URTICACEÆ.

Monœcious, diœcious, or barely polygamous herbs, shrubs, or trees, with stipules, and a regular calyx, free from the ovary, which forms a one-seeded fruit. Divides into four distinct subfamilies which might be reckoned as families, viz.: —

I. ELM SUBFAMILY. Trees, with alternate simple leaves, and polygamous or often nearly perfect flowers: styles or long stigmas 2.

Ovary 2-celled, a hanging ovule in each cell: stamens 4 to 9. Flowers earlier than the leaves. Fruit a thin key, winged all round, one-seeded (Fig. 207), (*Ulmus*) ELM.

Ovary one-celled, with one hanging ovule: stamens 5 or 6. Fruit a small drupe. Leaves ovate or heart-shaped, (*Celtis*) HACKBERRY.

II. BREADFRUIT SUBFAMILY. Trees, with a milky or colored juice, and alternate leaves; the flowers in heads or catkin-like spikes, the fertile ones fleshy in fruit, or both kinds in a fleshy receptacle. Styles 1 or 2: ovary becoming an akene in fruit. Inner bark often tough and fibrous.

Flowers, of both kinds mixed, enclosed in a pear-shaped fleshy receptacle like a rose-hip which is pulpy when ripe, (*Ficus*) *FIG.

Flowers monœcious, both kinds in separate catkin-like spikes; the calyx, &c. in the fertile sort becoming fleshy and eatable, making a berried multiple fruit (248,
Fig. 223). Stamens 4. Styles 2,   (*Morus*) MULBERRY.
Flowers diœcious: the fertile ones collected in a close and round head which is fleshy in
fruit. Stamens 4. Style 1.
    Sterile flowers in spikes. Leaves round-ovate or heart-shaped, rough above, soft-
downy beneath, some of them palmately lobed,   (*Broussonètia*) *PAPER-MULBERRY.
    Sterile flowers in racemes. Leaves oblong, smooth above, entire; branchlets spiny,
                                                    (*Maclùra*) *OSAGE-ORANGE.

III. NETTLE SUBFAMILY. Herbs (in this country), with opposite or alternate leaves, a tough
fibrous bark, and a colorless juice. Flowers monœcious or diœcious, in spikes, racemes, &c., not in
catkins. Stamens of the same number as the sepals. Ovary one-celled, and style or stigma only one;
fruit an akene.

Plants beset with stinging bristles.
    Leaves opposite: sepals 4 in both kinds of flowers: stigma a little tuft,   (*Urtica*) NETTLE.
    Leaves alternate: sepals 5 in the sterile, 4 unequal or 2 in the fertile, flowers: stigma
awl-shaped,   (*Laportea*) WOOD-NETTLE.
Plants destitute of stinging hairs, and
    Very smooth: leaves opposite: sepals 3 or 4, separate: stigma a tuft, • (*Pilea*) CLEARWEED.
    Smooth or hairy: leaves often alternate: calyx in the fertile flowers a cup with a
narrow mouth enclosing the ovary.
       Stigma long and thread-shaped: flower-clusters naked, in spikes, (*Bœhmèria*) FALSE-NETTLE
       Stigma a little tuft: flowers in axillary cymes or clusters, accompanied by
leafy bracts,   (*Parietària*) PELLITORY.

IV. HEMP SUBFAMILY. Herbs, with diœcious flowers, a colorless juice, fibrous-tough bark, and
opposite, or sometimes alternate, palmately-lobed or compound roughish leaves. Sterile flowers in
compound racemes or panicles, with 5 sepals and 5 stamens. Fertile flowers crowded, and with only
one sepal, which embraces the ovary and akene: stigmas 2, long.

Herb erect, annual: leaves of 5 to 7 lance-shaped toothed leaflets.. Stamens drooping.
    Fertile flowers in spiked clusters, each with a narrow bract,   (*Cànnabis*) HEMP.
Herb twining: root perennial: leaves heart-shaped and lobed. Fertile flowers in short
and scaly catkins, with broad and thin bracts, in fruit making a sort of
membranaceous cone,   (*Hùmulus*) HOP.

83. PLANE-TREE FAMILY. Order PLATANACEÆ.

This consists only of the genus

Plane-Tree. *Plátanus.*

Flowers monœcious, in separate round catkin-like heads. No calyx nor corolla to either kind.
Sterile flowers consisting of short stamens and club-shaped scales intermixed: fertile flowers, of little
scales and ovaries, which become club-shaped akenes, covered below with long hairs. Style awl-
shaped, simple. Trees, with colorless juice, alternate palmately-lobed leaves and sheathing stipules.
Only one species in this country, viz. :—

AMERICAN P., SYCAMORE, or BUTTONWOOD. A well-known tree by river-banks.   *P. occidentàlis.*

84. WALNUT FAMILY. Order JUGLANDACEÆ.

Timber and nut trees, with alternate pinnate leaves, no stipules; the sterile flowers in hanging catkins and with an irregular calyx; the fertile ones single or few together at the end of a shoot; their calyx coherent with the ovary, and 4-toothed at its summit. Fruit a kind of stone-fruit; the outer part becoming dry when ripe, and forming a husk, the stone incompletely 2-celled or 4-celled, but with only one ovule and seed. The whole kernel is a great embryo, with the cotyledons separated, lobed, and crumpled. — Only two genera : —

Catkins of the sterile flowers single; the bracts or scales united with the calyx: stamens 8 to 40. Fertile flowers with 4 small petals between the teeth of the calyx: short styles and stigmas 2, fringed: husk of the fruit thin, and not separating into valves or regular pieces. Bark and bruised leaves strong-scented and staining brown. Leaf-buds nearly naked, (*Juglans*) WALNUT.

Catkins 3 or more on one peduncle: stamens 3 to 8; anthers almost sessile. No petals in the fertile flowers: stigma large, 4-lobed. Husk of the fruit splitting into four pieces or valves, which separate from the smooth stone or shell. Wood very hard and tough. Leaf-buds scaly (Fig. 55), (*Cárya*) HICKORY.

Walnut. *Juglans.*

1. BLACK WALNUT. Leaves and stalks smoothish; leaflets many, lance-ovate, taper-pointed; fruit round, the thin husk drying on the very rough stone. Common W. *J. nigra.*

2. BUTTERNUT, or GRAY-BARKED W. Leaves, stalks, and oblong fruit clammy-downy when young, the stone with more ragged ridges, and tree smaller than No. 1. *J. cinèrea.*

3. TRUE or ENGLISH W. Smooth; leaflets only about 9. oblong; fruit round; husk separating from the thin and nearly smooth stone. Cultivated, from the South of Europe. *J. règia.*

Hickory. *Cárya.*

* Fruit and stone round or roundish.

1. SHAGBARK H. (also called SHELLBARK or SWEET H.) Bark on the trunk shaggy and scaling off; leaflets generally 5, three of them lance-obovate, the lower pair smaller and oblong-lanceolate, finely serrate; husk thick; stone roundish, thick or thin; seed very sweet: furnishes the hickory-nuts of the market. *C. alba.*

2. MOCKERNUT H. Bark cracked on the larger trunks ; leaflets 7 to 9, roughish-downy beneath, slightly serrate, oblong-lanceolate; catkins hairy; husk and stone very thick; seed sweetish but small. Common S. and W. *C. tomentòsa.*

3. PIGNUT H. Bark close and smooth; leaflets 5 to 7, smooth, lance-ovate, serrate; fruit pear-shaped or obovate, the husk and stone rather thin; seed sweetish or bitterish, small. *C. glabra.*

4. BITTERNUT or SWAMP H. Bark of trunk smooth; buds little scaly: leaflets 7 to 11, lance-oblong, smooth; husk and stone of the fruit thin and tender; seed very bitter. Wet woods. *C. amàra.*

* * Fruit and thin stone narrowly oblong: husk thin.

5. PECAN-NUT H. Leaflets 13 or 15, oblong-lanceolate, oblique, serrate; stone olive-shaped, thin; seed very sweet. W. & S. *C. olivæfórnuis.*

85. OAK FAMILY. Order CUPULIFERÆ.

Trees or shrubs, with alternate and simple straight-veined leaves, deciduous stipules, and monœcious flowers; the sterile flowers in slender catkins (or in head-like clusters in the Beech); the fertile flowers surrounded with an involucre which forms a cup, bur, or bag around the nut.

Fertile flowers scattered, or 2 or 3 together, their
  Involucre one-flowered, of many little scales, forming a cup around the base of the
      hard and roundish nut or acorn (Fig. 205),   (*Quercus*) OAK.
  Involucre containing 2 or 3 flowers, becoming a very prickly and closed bur enclos-
      ing the nuts, and splitting into 4 thick pieces.
    Nuts 1 to 3, roundish or flattish, thin-shelled. Sterile catkins long,   (*Castánea*) CHESTNUT.
    Nuts 2, sharply 3-angled. Sterile catkins like a head-like cluster,   (*Fagus*) BEECH.
  Involucre a leafy cup, lobed or torn at the end, longer than the bony nut, (*Córylus*) HAZEL.
Fertile flowers also collected in a kind of catkin. Nut small like an akene.
  Involucre an open 3-lobed leaf, 2-flowered,   (*Carpinus*) HORNBEAM.
  Involucre a closed bladdery bag, one-flowered, the whole catkin making a fruit like
      a hop in general appearance,   (*Ostrya*) HOP-HORNBEAM.

Oak. *Quercus.*

* Acorn ripening the first year, therefore borne on shoots of the season: cups stalked, except in No. 2: kernel generally sweet-tasted.

1. OVERCUP or BUR OAK. Leaves obovate, sinuate-pinnatifid, whitish-downy beneath; acorn 1' or 1½' long, in a deep cup with a mossy-fringed border.   *Q. macrocárpa.*

2. POST OAK. Leaves oblong, pale and rough above, grayish-downy beneath, pinnatifid, with 5 to 7 blunt lobes; cup saucer-shaped, much shorter than the acorn. Small tree.   *Q. obtusíloba.*

3. WHITE OAK. Leaves smooth when full grown, pale beneath, pinnatifid; the lobes 5 to 9, oblong or linear, entire; cup much shorter than the oval or oblong acorn. Rich woods.   *Q. alba.*

4. SWAMP CHESTNUT-OAK. Leaves obovate, whitish-downy beneath, coarsely and bluntly toothed or sinuate; cup thick, hemispherical, with stout or pointed scales; acorn oval, 1' long. *Q. Prinus.*

5. YELLOW CHESTNUT-OAK. Leaves lance-oblong, or oblong, acute, whitish, but scarcely downy beneath, rather sharply and evenly toothed; cup thin, and acorn smaller than in No. 4. Rich woods.   *Q. Castánea.*

6. CHINQUAPIN OAK. Much like No. 4, but a mere shrub, 2° to 6° high, with a thin cup and a smaller acorn. Sandy, barren soil.   *Q. prinoídes.*

* * Acorn ripening in the autumn of the second year; ripe fruit therefore on wood two years old, sessile: kernel bitter.

    ← Leaves entire or nearly so, narrow.

7. LIVE OAK. Leaves thick, evergreen, hoary beneath, oblong, small. Sea-coast, S.   *Q. virens.*

8. WILLOW OAK. Leaves light green, smooth, lance-linear, tapering, 3' or 4' long. S. & W. *Q. Phéllos.*

9. SHINGLE or LAUREL OAK. Leaves shining above, rather downy beneath, lance-oblong, thickish; cup saucer-shaped; acorn globular. Common S. & W.   *Q. imbricária.*

⁎ ⁎ Leaves or some of them a little lobed, broader upwards.

10. WATER OAK. Leaves smooth and shining, spatulate or wedge-obovate, with a tapering base; cup very short; acorn globular. Swamps, S.   *Q. aquática.*

11. BLACK-JACK OAK. Leaves thick and large, broadly wedge-shaped, and with 3 or 5 obscure lobes at the summit, shining above, rusty-downy beneath, the lobes or teeth bristle-pointed. Small tree, in barrens.   *Q. nigra.*

⁎ ⁎ ⁎ Leaves pinnatifid or lobed, long-stalked, the lobes or teeth bristle-pointed.

12. BEAR or SCRUB OAK. Leaves wedge-obovate, slightly about 5-lobed, whitish-downy beneath. A crooked shrub, 3° to 8° high; in barrens and rocky woods.   *Q. ilicifòlia.*

13. SPANISH OAK. Leaves grayish-downy beneath, narrow above, and with 3 to 5 irregular and narrow often curved lobes; acorn very short. Dry soil, S. & E. A fine tree.   *Q. falcàta.*

14. QUERCITRON OAK. Leaves rusty-downy when young, becoming nearly smooth when old, oblong-obovate, sinuate-pinnatifid; cup top-shaped, coarse-scaly; acorn globular or depressed. Large tree; the inner bark thick and yellow, used for dyeing.   *Q. tinctòria.*

15. SCARLET OAK. Very like the last, but the oval or oblong leaves smooth and shining, deeply pinnatifid (turning deep scarlet in autumn), the lobes cut-toothed; acorn rather longer than wide. Large tree, common in rich woods.   *Q. coccínea.*

16. RED OAK. Leaves smooth, pale beneath, oblong or rather obovate, with 4 to 6 short lobes on each side; acorn oblong-oval, 1' long, with a short saucer-shaped cup of fine scales. Common tree in rocky woods, &c.   *Q. rùbra.*

17. PIN or SWAMP SPANISH OAK. Leaves smooth and bright green on both sides, deeply pinnatifid, oblong ; the lobes diverging, cut and toothed, acute; acorn globular, only ½' long. Low grounds, N.   *Q. palústris.*

86. BIRCH FAMILY. Order BETULACEÆ.

Monœcious trees, with simple serrate leaves, and both kinds of flowers in scaly catkins (Fig. 146), two or three blossoms under each scale. Sterile flowers each with 4 stamens and a small calyx : fertile flowers with a 2 celled ovary bearing 2 long stigmas, and in fruit becoming a scale-like akene or small key. Only two genera : —

Sterile flowers with a calyx of one scale: fertile flowers 3 under each 3-lobed bract; each consisting of a naked ovary, in fruit becoming a broad-winged little key. Bark and twigs aromatic,   (*Bétula*) BIRCH.

Sterile flowers generally with a 4-parted calyx: fertile catkins short and thick, with hard scales, not falling off: fruit generally wingless,   (*Alnus*) ALDER.

Birch. *Bétula.*

1. WHITE BIRCH. A small and slender tree, with white outer bark; leaves triangular, very taper-pointed, on long and slender stalks. Common E.   *B. alba.*

2. PAPER B. A large tree, with white outer bark, peeling off in papery layers, and ovate or heart-shaped leaves. Common N.   *B. papyràcea.*

3. RIVER B. Tree, with ovate and angled acutish leaves, on short stalks, a brownish close bark, and short woolly fertile catkins. Common S. & W.   *B. nigra.*

CHERRY or SWEET B. Tree, with heart-ovate and pointed leaves, downy on the veins beneath, and a close bark, bronze-colored on the twigs, which are spicy-tasted, like the foliage of Checkerberry. Common N. *B. lenta.*

87. SWEET-GALE FAMILY. Order MYRICACEÆ.

Shrubs (generally low), with fragrant alternate leaves; and with catkins much as in the Birch family, but short and with only one naked blossom under each scale; the ovary forming a little nut or dry drupe.

Flowers monoecious: fertile catkins round and bur-like: fruit a smooth little nut. Leaves lance-linear, pinnatifid. Fern-like, whence the common name, (*Comptonia*) SWEET-FERN.

Flowers dioecious: scales of the fertile catkins falling off, and leaving only the small round fruits, which are incrusted with wax, and so appear like drupes. Leaves entire or serrate, (*Myrica*).

One species in wet grounds, N., with wedge-lanceolate pale leaves, (*M. Gale*) SWEET-GALE.
One on the sea-coast with lance-oblong, shining leaves, and waxy fruit, (*M. cerifera*) BAYBERRY.

88. WILLOW FAMILY. Order SALICACEÆ.

Dioecious trees or shrubs, with both kinds of blossoms in catkins (often earlier than the foliage); the flowers naked (without any calyx or corolla), one sort of two or more stamens under a scaly bract; the other of a one-celled pistil with two styles or stigmas, making a many-seeded pod: the seeds bearing a long tuft of down. Leaves alternate and simple: wood soft and light: bark bitter. — The Willows are of very many species, and are much too difficult for the beginner.

494. Shoot and catkin of sterile flowers of the Common White Willow. 495. A scale separated, with its flower, consisting of two stamens and a little gland, magnified. 496. Shoot and fertile catkin of the same. 497. A pistillate flower with its scale and gland, magnified.

Scales of the catkins entire: stamens 2 to 6: stigmas short: leaves narrow, (*Salix*) WILLOW.
Scales of the catkins cut-lobed: stamens 8 to 40: stigmas long: leaves broad. Scaly leaf-buds covered with a resinous varnish, (*Populus*) POPLAR.

89. PINE FAMILY. Order CONIFERÆ.

The only familiar family of Gymnospermous plants (218, 250), consisting of trees or shrubs, with resinous juice, mostly awl-shaped or needle-shaped leaves, and monœcious or diœcious flowers of a very simple sort, and collected in catkins, except in Yew. In that the fertile flower is single at the end of the branch. No calyx nor corolla, and no proper pistil. Ovules and seeds naked. Sterile flowers of a few stamens or anthers, fixed to a scale. Cotyledons often more than one pair, sometimes as many as 9 or 12, in a whorl. — For illustrations, see Fig. 49, 50, 134, 196, 197, 224 to 226, and 498, 499. — This family comprises some of our most important timber-trees, and the principal evergreen forest-trees of Northern climates. It consists of three well-marked subfamilies : —

498. Fertile flowers, or young cone, of Arbor-Vitæ, enlarged. 499 Inside view of one of the scales and its pair of naked ovules, more magnified.

I. PINE Subfamily. Fertile flowers many in a catkin, which in fruit becomes a *strobile* or cone (250); the scales of which are open pistils (each in the axil of a bract), with a pair of ovules or seeds borne on the base of each. Seeds scaling off with a wing. Cones ovate or oblong. Leaf-buds scaly. Flowers monœcious.

Leaves 2 to 5 in a cluster, from the axil of a thin scale, evergreen, needle-shaped. Cone
 with thick or sometimes thin scales, (*Pinus*) PINE.
Leaves many in a cluster (Fig. 134) on side spurs, and also scattered along the shoots of
 the season, needle-shaped, falling in autumn. Cone with thin scales, (*Larix*) LARCH.
Leaves all scattered along the shoots, evergreen, linear or needle-shaped. Cone with thin
 scales, (*Abies*) FIR.

II. CYPRESS Subfamily. Fertile flowers few, in a rounded catkin, formed of scales which are generally thickened at the top, and without any bracts, bearing one or more ovules at the bottom. Leaves scale-like or awl-shaped. Leaf-buds without any scales. Flowers monœcious. Cone dry, opening at maturity.

 Leaves deciduous and delicate, linear, 2-ranked. Cone round and woody, each shield-
  shaped scale 2-seeded, (*Taxòdium*) BALD-CYPRESS.
 Leaves evergreen, small, scale-like and awl-shaped (of two shapes).
  Cone woody and round; the scales shield-shaped, (*Cupréssus*) CYPRESS.*
  Cone of a few oblong and nearly flat loose scales (Fig. 498), (*Thuja*) ARBOR-VITÆ.*
Flowers diœcious, or sometimes monœcious. Fruit composed of a few closed scales,
  which become pulpy and form a sort of false berry, (*Juniperus*) JUNIPER.

III. YEW Subfamily. Buds scaly: leaves linear. Fertile flower single at the end of a branch, ripening into a nut-like seed. This is enclosed in an open and at length pulpy, berry-like red cup, in our only genus, viz. (*Taxus*) YEW.

* Our only *Cupressus* is *C. thyoides*, the WHITE CEDAR, rather common South. The ARBOR-VITÆ, *Thuja occidentalis*, so common North, and cultivated for evergreen hedges, is also called WHITE CEDAR. Our RED CEDAR is a Juniper.

Pine. *Pinus.*

Leaves 2 or 3 in a sheath, rigid: bark of tree rough: scales of the cones woody, thickened on the back at the end, and commonly tipped with a prickly point.

1. JERSEY or SCRUB PINE. Leaves in twos, only about 2' long. A straggling tree, S. & E. *P. inops.*
2. RED PINE (wrongly called *Norway Pine*); leaves in twos, 5' or 6' long; scales of the cones not pointed. A large tree, N. *P. resinòsa.*
3. YELLOW PINE. Leaves slender, in twos or threes, 3' to 5' long; cones small, their scales tipped with a weak prickly point. *P. mitis.*
4. PITCH PINE. Leaves rigid, dark green, in threes, 3' to 5' long; cones with a stout prickly point (Fig. 224). Common N. *P. rigida.*
5. LOBLOLLY PINE. Leaves in threes, 6' to 10' long, light green; cones 3' to 5' long. Light or exhausted soil. S. *P. Tæda.*
6. LONG-LEAVED PINE. Leaves in threes, 8' to 11' long, dark green; cones 6' to 8' long. Common S. & E. *P. austràlis.*

* * Leaves 5 together, slender: bark of young tree smooth: scales of cone naked and not thickened.

7. WHITE PINE. Leaves pale green; cones narrow, 4' or 5' long, hanging. A large tree, in moist woods North, with soft light wood. *P. Strobus.*

Larch. *Larix.*

1. AMERICAN LARCH or TAMARACK. Leaves very slender, short; cones not over 1' long, of few rounded scales. Swamps, N. *L. Americàna.*
2. EUROPEAN LARCH. A cultivated tree, with longer leaves and much larger cones than our wild species, the scales three times as many. *L. Europæa.*

Fir or Spruce. *Àbies.*

* Cones upright on short side-shoots, falling into pieces when ripe, the scales separating from the axis; leaves flat, becoming more or less 2-ranked, whitish beneath.

1. BALSAM FIR. Leaves narrowly linear; cones cylindrical, 3' or 4' long, 1' thick, bluish. Damp woods and swamps, N. *A. balsàmea.*

* * Cones hanging from the ends of branches, not falling to pieces.

2. HEMLOCK SPRUCE. Leaves linear, flat, ½' long, 2-ranked; cones oval, ¾' long. Hills. *A. Canadénsis.*
3. BLACK SPRUCE. Leaves needle-shaped, 4-sided, not 2-ranked, uniformly green; cones ovate, 1' to 1½' long, with thin edged scales. Swamps and cold woods. *A. nigra.*
4. WHITE or SINGLE SPRUCE. Cones oblong-cylindrical, 1' or 2' long, the scales with thickish edges; otherwise nearly like the last: found only at the North. *A. alba.*
5. NORWAY SPRUCE. Cones cylindrical, 5' to 7' long; leaves longer than in our wild species. A handsomer tree, from Europe, now commonly planted as an evergreen. *A. excélsa.*

Juniper. *Juniperus.*

1. COMMON JUNIPER. Shrub spreading; leaves in whorls of three, linear-awl-shaped, prickly-pointed, green beneath, white above; berries dark purple. Dry hills, N. *J. commùnis.*
2. SAVIN J. or RED CEDAR. Shrub or tree; leaves small and much crowded, awl-shaped and loose on vigorous shoots; on others smaller, scale-like, and closely overlying each other in 4 ranks; berries purplish with a white bloom. Dry hills. Wood reddish, very durable. *J. Virginiàna.*

CLASS II. — ENDOGENS OR MONOCOTYLEDONS.

Stem having the wood in threads or bundles, interspersed among the pith or cellular part, not forming a ring or layer, and not increasing by annual layers.

Leaves parallel-veined, not branching and forming meshes of network. To this some Arums, Trillium, Greenbrier, &c. are exceptions, having more or less netted veins.

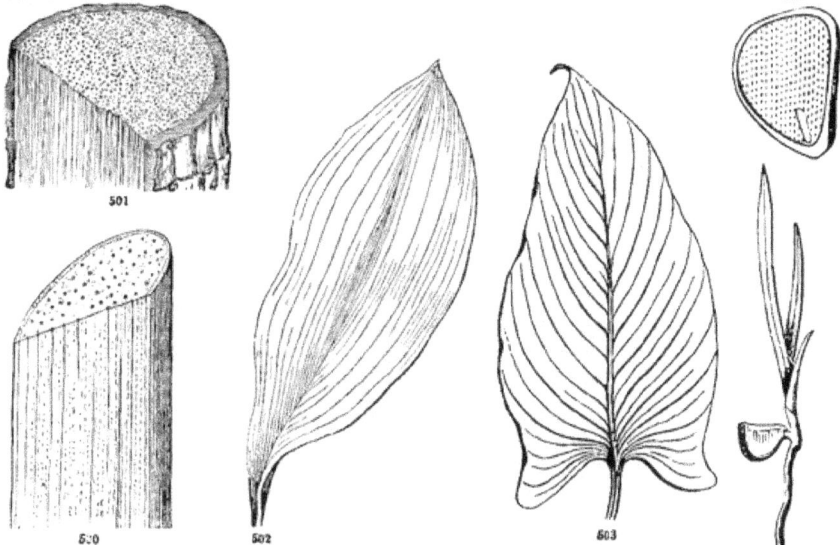

500. Endogenous stem of one year old, shown in a Corn-stalk.   501. One of several years old, of Palmetto. Parallel-veined leaves of the two kinds: 502. that of Lily of the Valley ; 503. one of Calla   504. Magnified section of the seed of Iris, showing the small monocotyledonous embryo.   505. Plantlet of Iris growing from the seed.

Flowers with their parts mostly three or six, never five.

Embryo monocotyledonous, i. e. of only one true seed-leaf: so in germination the leaves are all alternate or one above another.

Except the Palmetto and one or two Yuccas at the South (Fig. 79), and some Greenbriers, all the Endogens of this country are herbs. In warmer climates there are many Palms and other woody plants of the class, all having an appearance very different from our common trees and shrubs (113, 114).

KEY TO THE FAMILIES OR ORDERS OF CLASS II.

I. Spadiceous Division. Flowers collected on a spadix (184), i. e. sessile and crowded in a spike or head on a thickened axis, and with or without a spathe or enwrapping bract (185).

Trees or shrubs, with simple stems; the flowers having calyx and corolla, PALM FAMILY, 205
Herbs, the small and crowded flowers either naked or with a small perianth.
　　Spadix surrounded by a large spathe: flowers generally naked: fruit a berry, }
　　Spadix without a spathe: perianth of 6 pieces, } ARUM F. 205
　　Spadix without any proper spathe: perianth none: fruit an akene, CAT-TAIL F. 206
　　Spadix (as it might be called) raised above a small spathe, covered with blue and
　　　　tubular, 6-lobed flowers. Belongs to the next division, PICKEREL-WEED F. 208

II. Petaloideous Division. Flowers not on a spadix and not enclosed by glumes or chaffy or scale-like bracts (as in Grasses and Sedges), but having a calyx and corolla, or a 6-leaved or 6-lobed (rarely 4-leaved) perianth colored like a corolla.

Perianth free from the ovary, that is, inserted underneath the ovary, and
　　Of 3 green or greenish sepals and 3 distinct and colored petals.
　　　　Pistils many, in a ring or a head, making akenes, WATER-PLANTAIN F. 206
　　　　Pistil 1: styles or sessile stigmas 3. Leaves whorled, veiny, TRILLIUM F. 206
　　　　Pistil and slender style 1: leaves alternate, parallel-veined, SPIDERWORT F. 207
　　Of mostly 6 petal-like leaves in two ranks, three outside and three inside, or else 6-
　　　　(rarely 4-) lobed, all colored alike.
　　　　Stamens only 3, or 6 and the three on one side of the flower much shorter than
　　　　　　the rest, PICKEREL-WEED F. 208
　　　　Stamens 6, or as many as the divisions of the perianth, all alike.
　　　　　　Anthers turned outwards, i. e. on the outer side of the filament.
　　　　　　　　Leaves in whorls: flowers perfect: long stigmas 3, INDIAN CUCUMBER-ROOT, 207
　　　　　　　　Leaves alternate, and with side tendrils, netted-veined between the ribs:
　　　　　　　　　　flowers dioecious: styles or sessile stigmas 3, GREENBRIER F. 208
　　　　　　　　Leaves alternate, without tendrils: flowers perfect or polygamous: styles
　　　　　　　　　　3 or 3-cleft, COLCHICUM F. 209
　　　　　　Anthers turned inwards, i. e. on the inner side of the filament: style 1:
　　　　　　　　stigmas 1 or 3, LILY F. 209
Perianth adherent to the ovary below, and therefore apparently borne on it.
　　Stamens 6: anthers turned inwards. Flowers regular or nearly so, AMARYLLIS F. 213
　　Stamens 3: anthers turned outwards. Flowers often irregular, IRIS F. 214
　　Stamens only one or two and united with or borne on the style. Flowers irregular, of
　　　　singular shapes, ORCHIS F. 215

III. Glumaceous Division. Flowers not on a spadix, and without any corolla-like perianth, but with *glumes*, i. e. thin scales, such as the chaff or husk of Grain and Grasses. Stems rush-like or straw-like.

Glumes 6 in a whorl to each flower, like a calyx, RUSH F. 215
Glume one to each flower, the flower in its axil: Flowers collected into heads or spikes, SEDGE F. 216
Glumes 2 or 4 to each flower, in two sets, GRASS F. 216

I. Spadiceous Division.

90. PALM FAMILY. Order PALMÆ.

Although some, like the Dwarf Palmettos of the Southern States, make only rootstocks not rising out of the ground, most Palms form trees, with a simple, unbranched, cylindrical trunk, growing by the terminal bud only, and always surmounted by a crown of large and peculiar, long-petioled leaves. These are fan-shaped in the Palmetto (Fig. 79), pinnate in the Date-Palm, &c. The flowers burst forth from a spathe; are small, but generally perfect, and furnished with a perianth of 6 parts, in two sets, the outer answering to a calyx, the inner to a corolla. Fruit a nut; that of the Cocoanut is a good illustration. The principal Palms of our southern sea-coast belong to the genus (*Chámærops*) PALMETTO.

91. ARUM FAMILY. Order ARACEÆ.

Herbs with sharp-tasted or acrid juice, and more or less fleshy in their texture; the leaves either simple or compound, and commonly so much netted-veined that the plants might readily be mistaken for Exogens. The small flowers are closely spiked or packed on a fleshy axis, forming a spadix. The fruit is a berry, or sometimes dry and leathery, but containing some pulp or jelly. The following are the principal genera we meet with.

Spathe present, forming a hood, wrapper, or a petal-like leaf.
    Flowers naked, i. e. without any perianth, monœcious, diœcious, or polygamous,
        Covering only the base of the long spadix, which is enclosed in the hooded
            spathe (Fig. 147). Stem simple, from a rounded corm: leaves com-
            pound, of 3 or more leaflets,   (*Arisæma*) INDIAN-TURNIP.
        Covering the whole length of the spadix. Leaves simple, arrow-shaped (Fig.
            503) or heart-shaped: spadix on the end of a scape, bearing stamens
            only at the upper part.
        Spathe green, thick, and closely folded around the spadix: anthers sessile.
            Herb growing in shallow water,   (*Peltándra*) ARROW-ARUM.
        Spathe white and petal-like, open,   (*Cálla*) CALLA.
    Flowers with a 4-leaved perianth or calyx, perfect, on a globular spadix, surrounded
            by a thick, shell-shaped, purplish spadix coming out of ground in
            earliest spring, some time before the great ovate and heart-shaped, veiny
            leaves; odor that of the skunk. Stamens 4,   (*Symplocárpus*) SKUNK-CABBAGE.
Spathe none at all; the spadix naked, covered with flowers, which are perfect, with a
            perianth of 6 or sometimes 4 pieces, and as many stamens.
    Spadix on the summit of a scape rising out of the water: leaves oblong, on a long
            petiole,   (*Oróntium*) GOLDEN-CLUB.
    Spadix from the side of a leaf, or from a stem similar to one of the long and erect,
            linear, 2-edged or sword-shaped leaves: all springing from a sharp-
            aromatic and creeping rootstock,   (*Ácorus*) SWEET-FLAG.

92. CAT-TAIL FAMILY. Order TYPHACEÆ.

Marsh herbs, with linear, sword-shaped leaves (erect, except they float in water), and monœcious naked flowers in dense spikes or heads, one sort consisting of some stamens only, the other of pistils only. Fruit a one-seeded akene. No spathe, except some open bracts or leaves.

Flowers in one long spike or spadix, the upper part bearing stamens only, the lower slen-
      der pistils only closely packed together; ovary long-stalked and surrounded
      by slender down,   (*Typha*) CAT-TAIL.
Flowers in separate heads, some bearing stamens only, others pistils only, each sur-
      rounded by several scales, but no down,   (*Spargànium*) BUR-REED.

II. Petaloideous Division.

93. WATER-PLANTAIN FAMILY. Order ALISMACEÆ.

Marsh or aquatic herbs, with a distinct calyx of 3 green or greenish sepals and a corolla of 3 white petals, 6 to many stamens on the receptacle, and many one-ovuled pistils collected into a ring or head, becoming akenes in fruit. Leaves mostly oblong-heart-shaped, lance-shaped, or arrow-shaped, sometimes with cross veinlets, long-petioled. Flowers on scapes. Two genera are common.

Flowers perfect with about 6 stamens, small, in an open panicle: pistils 15 to 20 in a ring:
      leaves not arrow-shaped,   (*Alisma*) WATER-PLANTAIN.
Flowers monœcious or diœcious, in a loose raceme or spike; the sterile ones with many
      stamens; the fertile with many pistils in a head, making thin winged akenes.
      Leaves or some of them generally arrow-shaped,   (*Sagittària*) ARROWHEAD.

94. TRILLIUM FAMILY. Order TRILLIACEÆ.

Herbs with simple stems rising from a short rootstock, rather conspicuously netted-veined leaves in a whorl, and perfect and regular flowers : — containing in this country only the genus Trillium and the Indian Cucumber-root, which are here described.

Trillium.*   *Trillium.*

Stem bearing at the summit a whorl of 3 broad leaves and one rather large flower. Calyx of 3 green spreading sepals. Corolla of 3 spreading petals. Stamens 6, with short filaments and long erect anthers turned inwards, inserted on the receptacle. Pistil one, 3-celled, commonly with 3 to 6 lobes or ridges, and making a purple many-seeded berry in fruit: styles or long sessile stigmas 3, spreading. — They all grow in rich woods, and blossom in spring or early summer.

1. SESSILE FLOWERED T. Flower and the ovate leaves both sessile; petals rather erect, dark dull
      purple or greenish. W. & S.   *T. séssile.*
2. RECURVED T. Leaves narrowed at the base into a footstalk; sepals turned down; petals nar-
      rowed at both ends; otherwise like No. 1. W.   *T. recurvàtum.*

---

      * Also called BIRTHROOT, WAKE-ROBIN, and THREE-LEAVED NIGHTSHADE.

3. NODDING T. Leaves nearly sessile, rhombic-ovate ; flower small, on a short peduncle curved down under the leaves; petals oblong-ovate, pointed, recurved, wavy. E. & S.   *T. cérnuum.*

4. ERECT T. or BIRTHROOT. Leaves sessile, round-rhombic with a very abrupt point; flower on a nearly upright peduncle; petals ovate, acutish, spreading, dull purple or sometimes greenish-white. Common N.   *T. eréctum.*

5. GREAT-FLOWERED T. Leaves and peduncle nearly as in No. 4; petals obovate, erect at the base, then gradually spreading much longer and broader than the sepals, white, turning rose-color when old. N. and W.   *T. grandiflòrum.*

6. PAINTED T. Leaves petioled, pale green, ovate, taper-pointed; flower on an upright peduncle; petals lance-ovate, pointed, widely spreading, longer than the sepals, wavy, white, adorned with delicate pink-purple stripes at the base. Cold damp woods, &c. N.   *T. erythrocárpum.*

506. Flower of Trillium, natural size.

Indian Cucumber-root. *Medèola.*

Stem 1° to 3° high, from a white tuberous horizontal rootstock, having the taste of a cucumber, bearing near the middle a whorl of 5 to 9 obovate-lanceolate pointed sessile leaves, and at the top one of 3 ovate smaller leaves, and a few small greenish-yellow flowers in an umbel, on recurved stalks. Sepals and petals each 3, oblong and alike, recurved. Stamens 6: filaments longer than the anthers. Stigmas 3, sessile, long and thread-shaped. Ovary one, making a round 3-celled and few-seeded berry. One species, in damp woods; flowering in summer.   *M. Virgínica.*

95. SPIDERWORT FAMILY. Order COMMELYNACEÆ.

Tender herbs, with alternate parallel-veined leaves sheathing at the base, and perfect flowers, having 3 green or greenish sepals and 3 petals on the receptacle. Pistil one, with one long style and one stigma. Pod small, 3-celled or sometimes 2-celled, few-seeded. Flowers opening in the morning for only one day, the delicate (generally blue or purple) petals then melting away. There are two genera wild; and the Spiderwort is cultivated in every flower-garden.

Flowers regular: the 3 petals and 6 stamens all alike: filaments bearded with jointed colored hairs : leaves lance-linear, sessile, all alike,   (*Tradescántia*) SPIDERWORT.

Flowers irregular: two of the petals kidney-shaped on long claws, and one smaller: stamens unequal, only three of them with good anthers : filaments naked : lower leaves with sheathing footstalks, the uppermost sessile and somewhat heart-shaped,   (*Commelýna*) DAY-FLOWER.

508   507

5.7 Flower of Spiderwort.
5.8 Pistil, magnified; the ovary cut across.

96. PICKEREL-WEED FAMILY. Order PONTEDERIACEÆ.

Is represented by three or four plants in this country, of which much the commonest is the

Pickerel-weed. *Pontedèria.*

Perianth blue, of 6 divisions, unequally united below into a tube ; the 3 upper divisions most united and making a 3-lobed upper lip, the 3 lower spreading and separate some way down, making a lower lip: after expanding, for one day only, the upper part coils up and withers away, while the base of the tube thickens and encloses the small one-seeded fruit. Stamens 6; the 3 lower on slender projecting filaments; the 3 upper inserted lower down on the tube, with very short filaments and generally imperfect anthers. Style 1 : stigma 3-lobed. Stout herbs in shallow water, with long-petioled leaves and long peduncles or few-leaved stems (their leaves with sheathing footstalks, the uppermost one merely a sheathing spathe or bract), bearing a spike of flowers.

1. COMMON PICKEREL-WEED. Stems 2° or 3° high; leaves thickish, lance-ovate or ovate-oblong, and generally more or less heart-shaped at the base. Common everywhere; fl. all summer. *P. cordàta.*

97. GREENBRIER FAMILY. Order SMILACEÆ.

Of this family, as here arranged, we have only a single genus, viz. : —

Greenbrier. *Smilax.*

Known at once by being climbing plants (or disposed to climb) and having a tendril on each side of the footstalk of the leaf; and by the leaves being veiny between the ribs, almost as in Exogens, alternate, sometimes evergreen, simple, and entire. Flowers diœcious. in axillary umbels. Perianth generally of 6 equal and spreading greenish or yellowish separate pieces. The sterile flowers have as many stamens, with oblong or linear one-celled anthers fixed by their base to the filament, and turned inwards. The fertile flowers have a round ovary, with 3 short spreading styles or stigmas. Fruit a berry, with 2 or few large seeds. Fl. summer.

* Stems woody and often prickly, yellowish-green: ovary and berry 2-celled and 2-seeded, black when ripe, generally with a bluish bloom.

1. COMMON G. or CATBRIER. Leaves thickish, round-ovate or slightly heart-shaped, and with 5 to 9 ribs, green both sides; branchlets often square; prickles short; peduncles of the umbel not longer than the petiole. Moist thickets. *S. rotundifòlia.*

2. GLAUCOUS G. Leaves ovate, glaucous beneath ; peduncles longer than the petiole : otherwise nearly as No. 1. *S. glauca.*

3. BRISTLY G. Leaves ovate and heart-shaped, large and thin, green both sides; stem below covered with long and weak blackish bristly prickles; peduncles much longer than the petioles. Thickets, N. and W. *S. híspida.*

4. LAUREL-LEAVED G. Not prickly; leaves lance-oblong or lance-linear, thick and evergreen, with 3 to 5 ribs; peduncles of the umbel very short. Sandy soil, S. *S. laurifòlia.*

   * * Stem herbaceous, climbing, not prickly: ovary and blue-black berry 3-celled, 6-seeded.

5. CARRION-FLOWER G. Leaves thin, pale, mostly heart-shaped, with 7 to 9 ribs, sometimes rather downy beneath, long-petioled; peduncles 3' to 8' long, longer than the leaves; flowers of the odor of carrion. Meadows and river-banks. *S. herbàcea.*

98. COLCHICUM FAMILY. Order MELANTHACEÆ.

Herbs, with parallel-veined leaves; the flowers generally perfect or polygamous; the perianth of 6 similar divisions colored alike; the 6 stamens with their anthers turned outwards. Ovary one, 3-celled, bearing 3 styles, which are generally separate, but sometimes united into one. Many are acrid or poisonous plants, none more so than the common Veratrum or White-Hellebore, which is often called *Poke*, a name which properly belongs to Phytolacca, p. 191.

Flower and leaves rising from a corm underground: perianth a long tube, bearing 6 similar petal-like lobes, (*Colchicum*) *COLCHICUM.
Flowers with a perianth of 6 separate leaves.
    Perianth persisting or withering without falling. Plants acrid-poisonous: flowers polygamous, in panicles, terminating the simple leafy stem.
        Divisions of the perianth on claws, bearing the stamens: leaves narrow: flowers cream-colored, turning greenish-brown with age, (*Melanthium*) MELANTHIUM.
        Divisions of the perianth without claws, greenish. Leaves oval or oblong, partly clasping, plaited, (*Veratrum*) WHITE-HELLEBORE.
    Perianth falling off after flowering. Plants not poisonous: stems generally forking: leaves sessile or clasping, ovate or lance-oblong: flowers perfect, generally single, nodding: divisions of the perianth long and narrow.
        Styles united into one at the bottom. Perianth large, lily-like, yellowish: fruit a few-seeded pod. Flower-stalk not twisted or jointed. (*Uvularia*) BELLWORT.
        Styles united into one almost to the top. Divisions of the whitish or rose-colored perianth recurved: fruit a many-seeded red berry. Flower-stalks single in the axil of the leaves, and with a joint or abrupt bend or twist in the middle, (*Streptopus*) TWIST-STALK.

Bellwort. Uvularia.

1. LARGE-FLOWERED B. Leaves oblong, clasping-perfoliate, i. e. the stem appearing to run through the lower part of the leaf; perianth pale greenish-yellow, 1½' long. Rich woods, N. & W. (All the species flower in the spring.) *U. grandiflora.*

2. PERFOLIATE B. Like the last, but the flower smaller and yellow, and the anthers more pointed. Common E. *U. perfoliata.*

3. SESSILE-LEAVED B. Smaller than the rest; leaves sessile, not encompassing the stem; flower cream-color. *U. sessilifolia.*

99. LILY FAMILY. Order LILIACEÆ.

A large family, with much variety in appearance. Leaves parallel-veined, and sessile or sheathing. Flowers perfect and regular; the perianth of 6 divisions or lobes (or in one case with only 4), all colored alike, inserted on the receptacle free from the ovary. Stamens as many as the parts of the perianth, with their anthers turned inwards. Pistil one, with a 3-celled (rarely 2-celled) ovary and a single style; but with as many stigmas, or lobes to the stigma, as there are cells in the ovary. Fruit a pod or a berry.

Fruit a few-seeded berry: flowers small. Herbs from rootstocks: no bulbs.

Stems much branched: leaves fine and thread-shaped, in clusters,   (*Aspáragus*) *ASPARAGUS

Stems simple above ground and leafy.   Leaves oblong or lance-oblong.

Flowers axillary, nodding, greenish; perianth tubular, 6-lobed: stamens above
the middle, on very short filaments.   Rootstock thick, marked with
broad round scars on the upper side (Fig. 63),   (*Polygonàtum*) SOLOMON'S-SEAL.

Flowers in a terminal raceme, white: perianth 6-parted, in one case 4-parted,
the divisions narrow and widely spreading, the stamens on its base:
filaments slender,   (*Smilacìna*) SMILACINA.

Stems or scape simple and leafless above ground; the broad leaves all from its base
or from the slender rootstock.

Flowers small, in a slender raceme, white; perianth bell-shaped, 6-lobed (Fig. 3):
leaves very smooth,   (*Convallària*) *LILY-OF-THE-VALLEY.

Flowers rather large, in an umbel, greenish-yellow or whitish: perianth 6-leaved:
leaves of the plant ciliate,   (*Clintònia*) CLINTONIA.

Fruit a 3-celled pod, splitting into 3 valves when ripe.

Perianth wheel-shaped, or sometimes erect or bell-shaped, 6-leaved: flowers on a
scape or nearly naked stem, rising from a coated bulb: seeds round
and black, few.

Flowers in a corymb, white: style 3-sided,   (*Ornithógalum*) *STAR-OF-BETHLEHEM.

Flowers in a raceme, blue or purple: style thread-like,   (*Scìlla*) SQUILL.

Flowers in an umbel from a scaly bract or involucre,   (*Állium*) ONION.

Perianth funnel-shaped, bell-shaped, or globe-shaped, more or less united into a tube
or cup, bearing the 6 stamens, except in some Day-Lilies.

Scape and leaves from a coated bulb: flowers in a raceme.   Leaves narrow.

Perianth globular, blue, small,   (*Muscàri*) *GRAPE-HYACINTH.

Perianth short, funnel-shaped or bell-shaped, 6-cleft,   (*Hyacìnthus*) *HYACINTH.

Scape or stem leafy towards the bottom, from fibrous roots (no bulb), bearing a
few large flowers in a cluster at the top: stamens curved to one side.

Flower opening for only one day,   (*Hemerocállis*) *DAY-LILY.

Perianth bell-shaped or funnel-shaped, &c., but of 6 separate petal-like divisions:
seeds many, mostly flat, pale.

Simple-stemmed herbs from a scaly or coated bulb: stamens on the receptacle
or attached to the very base of the deciduous perianth.

Anthers fixed by their middle and swinging free: stems leafy to the top.

No honey-bearing spots, or merely a groove at the bottom of each division of the perianth. Bulb scaly,   (*Lilium*) LILY.

A round and large honey-bearing spot near the bottom of each division
of the perianth,   (*Petilium*) *CROWN-IMPERIAL.

Anthers erect on the filament, appearing to be fixed by their base: stem or
scape leafy only at or towards the bottom.

Style none or hardly any: stigmas 3 on the long 3-sided ovary,   (*Tùlipa*) *TULIP.

Style long: ovary roundish: leaves 2, spotted,   (*Erythrònium*) DOGTOOTH-VIOLET.

Stems woody, palm-like, or not rising above the ground, from roots or rootstocks
(no bulbs): leaves evergreen, sword-shaped.   Flowers white, tulip-
shaped, in a large, terminal, compound panicle,   (*Yucca*) YUCCA

Smilacina (or FALSE SOLOMON'S-SEAL). *Smilacina.*

1. RACEMED S. Minutely downy, 2° or 3° high, many-leaved; leaves lance-oblong, tapering abruptly at both ends, ciliate; flowers many, in compound racemes. Moist grounds. *S. racemòsa.*

2. STAR-FLOWERED S. Nearly smooth, 1° or 2° high; leaves many, lance-oblong, slightly clasping, pale beneath; raceme simple and few-flowered. Moist thickets, &c., N. *S. stellàta.*

3. THREE-LEAVED S. Smooth, 3′ to 6′ high; leaves commonly 3, oblong, tapering into a sheathing base; flowers several, in a slender simple raceme. Bogs, N. *S. trifòlia.*

4. TWO-LEAVED S. Nearly smooth, 3′ to 5′ high, with commonly 2 heart-shaped leaves, the lower one generally petioled; flowers in a simple short raceme; perianth 4-parted, reflexed; stamens 4. Moist woods, in spring. *S. bifòlia.*

Onion (GARLIC and LEEK). *Állium.*

§ 1. ONION proper, with hollow, stem-shaped leaves, and an open, widely spreading, star-shaped blossom.

1. GARDEN ONION. Scape naked, much longer than the leaves, hollow, swollen in the middle; flowers whitish; umbel often bearing small bulbs (top-onions); the large bulb turnip-shaped. Commonly cultivated. *A. Cepa.*

2. CHIVES O. Scape naked, about as long as the slender leaves; all growing in tufts, from small bulbs; flowers purplish, crowded. Cultivated. *A. Schœnoprásum.*

§ 2. GARLICS and LEEKS. Leaves flat or keeled and not hollow, except in No. 3.

3. FIELD GARLIC. Leaves thread-shaped, slender, round, but channelled on the upper side, hollow; bulbs small; umbel bearing flowers with a green-purple erectish perianth, or else only bulblets. Naturalized in low pastures and gardens. *A. vineàle.*

4. TRUE or ENGLISH GARLIC. Bulbs clustered and compound; leaves lance-linear, nearly flat; umbel bearing pale purple flowers with an erectish perianth, or else bulblets. Cultivated in gardens; not common. *A. sativum.*

5. GARDEN LEEK. Bulb single; leaves linear-oblong, acute, somewhat folded or keeled; flowers crowded in the umbel; perianth erectish, violet-purple. Rarely cultivated. *A. Porrum.*

6. WILD LEEK. Bulbs clustered, narrow, oblong, and pointed; leaves lance-oblong, blunt, flat, dying off by midsummer, when the naked scape appears with its loose umbel of white flowers; pod 3-lobed. Rich woods, N. and W. *A. tricòccum.*

Day-Lily. *Hemerocállis.*

✱ Flowering stems tall, leafy towards the bottom, somewhat branched above: leaves long and linear, keeled, 2-ranked: stamens on the top of the narrow tube of the perianth: seeds black and wingless.

1. COMMON DAY-LILY. Flower dull orange-yellow; inner divisions wavy, blunt. Gardens. *H. fulva.*

2. YELLOW D. Flower light yellow; inner divisions of the perianth acute. Gardens. *H. flava.*

✱ ✱ Flowering stems naked, simple: leaves broad and flat, ovate or oblong, and often heart-shaped, with veins springing from the midrib, long-stalked; stamens on the receptacle: seeds flat and winged (*Funkia*).

3. WHITE D. Flower white, funnel-shaped; leaves more or less heart-shaped. Gardens. *H. Japónica.*

4. BLUE D. Flower blue or bluish, the upper part more bell-shaped than in No. 3; leaves scarcely heart-shaped. Gardens. *H. cœrúlea.*

Lily. *Lilium.*
Foreign species, everywhere cultivated.

1. **WHITE LILY.** Leaves lance-shaped, scattered along the stem; flowers erect; perianth bell-shaped, white, smooth inside.   *L. album.*

2. **BULB-BEARING L.** Leaves lance-shaped, scattered along the tall stem, producing bulblets in their axils; flowers several, erect; perianth open-bell-shaped, orange-yellow, rough inside. *L. bulbiferum.*

        * * Wild species: flowers orange-colored, reddish, or yellow.

3. **WILD ORANGE L.** Stem 1° to 3° high, bearing scattered (or sometimes whorled) lance-linear leaves and 1 to 3 erect reddish-orange open-bell-shaped flowers, the 6 lance-shaped divisions narrowed at the base into claws, purplish-spotted inside.   Common in light or sandy soil.   *L. Philadelphicum.*

4. **WILD YELLOW L.** Stem 2° to 4° high, bearing distant whorls of lance-shaped leaves and a few nodding flowers on slender peduncles; perianth yellow or orange, with brown spots inside, bell-shaped with the divisions spreading or recurved to the middle.   Moist meadows, and along streams. (Fig. 1.)   *L. Canadénse.*

5. **SUPERB** or **TURK'S-CAP L.** Stem 4° to 7° high, only the lower leaves in whorls; flowers many, bright orange or reddish, with strong brown-purple spots inside, more recurved and larger than the last, but very much like it.   Rich low grounds.   *L. supérbum.*

Dogtooth Violet. *Erythrònium.*

1. **YELLOW D.** or **ADDER'S-TONGUE.** Leaves oblong-lance-shaped, pale-dotted, much blotched; flower pale yellow; style club-shaped, stout; stigmas united. Moist grounds : fl. in early spring.   *E. Americànum.*

2. **WHITE D.** Flower white or bluish; the style less thick than in No. 1. Rather common W.   *E. álbidum.*

3. **EUROPEAN D.** Leaves ovate or oblong, scarcely spotted; flowers purple or rose-color; style thread-shaped and not thickened upwards; stigmas separate.   Cultivated ; not common.   *E. Dens-cànis.*

509. Yellow Dogtooth-Violet.
510. The bulb.
511. Perianth laid open, and stamens.
512. The pistil, enlarged.
513. Lower half of a pod, cut across and magnified.

512
513
510   509   511

100. AMARYLLIS FAMILY. Order AMARYLLIDACEÆ.

Like the Lily Family, but with the (regular or slightly irregular) 6-cleft perianth coherent below with the surface of the ovary, and therefore in appearance inserted on its
summit. Stamens 6. Fruit a 3-celled pod. Herbs generally with naked stems or scapes,
and long linear leaves, from a coated bulb, commonly with showy flowers. Herbage and
bulbs acrid and poisonous.

Flower with a cup or crown at the throat of the salver-shaped or funnel-shaped perianth.
  Stamens long, from the edge of the cup-shaped crown: anthers linear, swinging free:
    divisions of the perianth long and narrow, recurved. Flowers white,
    showy; the cluster leafy-bracted,   (*Pancràtium*) *PANCRATIUM.
  Stamens included in the cup, unequal: filaments very short. Flowers from a scale-
    like spathe,   (*Narcissus*) *NARCISSUS.
Flower without any cup or crown on the perianth.
  Anthers fixed by the middle and swinging free, linear or oblong: filaments generally
    curved. Flowers large and showy, generally red or pink, (*Amaryllis*) *AMARYLLIS.
  Anthers erect on the filament.
    Flowers in a spike, funnel-shaped, white, very fragrant, (*Poliànthes*) *TUBEROSE.
    Flowers in an umbel, or single: perianth 6-parted down to the ovary.
      Flower single, from a 1-leaved spathe, white, nodding: three inner divisions
      of the perianth shorter than the three outer, and notched at the end:
      anthers long-pointed,   (*Galànthus*) *SNOWDROP
      Flowers one or more from a 1-leaved spathe, white, nodding; the 6 divisions
      of the perianth alike, often green-tipped: anthers blunt, (*Leucòium*) *SNOWFLAKE.
    Flowers few, with 2 small bracts at the base of the pedicels; the star-shaped
    perianth yellow, closing and remaining on the pod. Leaves grass-
    like, hairy. Plant small,   (*Hypòxys*) STAR-GRASS.

Narcissus. *Narcissus.*

Tube of the flower slender; the cup or crown much shorter than the 6 spreading divisions; anthers
  borne on the inside of the cup, or 3 of them a little protruding, on short filaments.

1. POET'S N. Scape flattish, tall, mostly one-flowered; flower white, the very short and flat crown
  yellow, generally margined with crimson or pink; sweet-scented; leaves bluntly keeled, rather
  glaucous. Gardens.   *N. poéticus.*

2. JONQUIL N. Flowers 1 to 4, on a round and slender scape, yellow, very fragrant, the cup saucer-
  shaped; leaves terete, channelled down one side. Gardens.   *N. Jonquilla.*

3. POLYANTHUS N. Flowers several, on a flattish scape, white, with a bell-shaped cup, not fragrant;
  leaves flat, glaucous. Gardens.   *N. Tazétta.*

* * Tube of the flower short, funnel-shaped; the cup or crown very large, bell-shaped, with a wavy-
crisped or toothed margin, equalling or longer than the 6 divisions of the perianth, and bearing
the stamens on its base.

4. DAFFODIL N. Flower one, large, sulphur-yellow, with a deeper yellow cup, on a flattened scape
1° high; leaves flattish. In all gardens; most common with flowers double, so that their structure
is obscured.   *N. Pseudo-Narcissus.*

101. IRIS FAMILY. Order IRIDACEÆ.

Herbs with perennial roots, commonly with rootstocks, bulbs, or corms, and with equitant leaves (151, Fig. 64); the flowers perfect, regular or irregular; tube of the corolla-like perianth below coherent with the surface of the ovary, and so appearing to grow from its summit; stamens only 3, one before each of the outer divisions of the perianth; their anthers turned outwards, i. e. looking towards the perianth and opening on that side. Ovary 3-celled, making a many-seeded pod: style one: stigmas 3, often flat or petal-like. Herbage, rootstocks, &c. generally acrid or sharp-tasted. Flowers generally showy, and from a spathe of one or more leaf-like bracts, or from the axils of the uppermost leaves, each one generally opening but once.

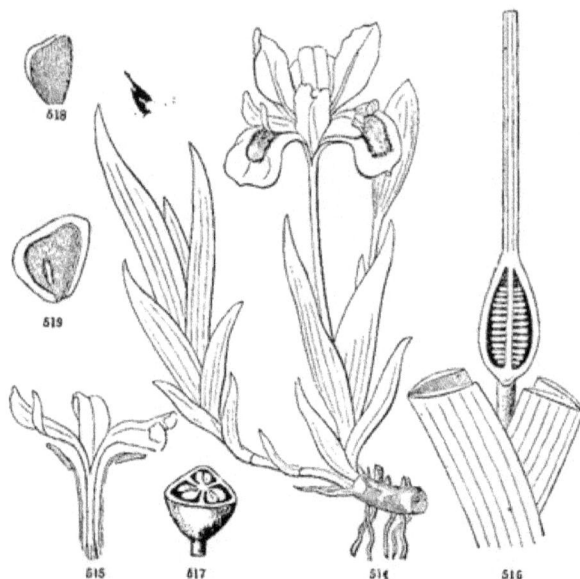

514. Plant of Crested Dwarf Iris 515. Top of the style and the 3 petal-like stigmas, also 2 of the stamens 51> Magnified pistil and lower part of the tube of the perianth, divided lengthwise: the folings cut away. 517. Lower part of a pod, divided crosswise. 518. Seed. 519. Magnified section of the same, showing the embryo

Filaments monadelphous in a tube which encloses the style as in a sheath: stigmas thread-shaped: perianth 6-parted nearly to the ovary, widely spreading, opening in sunshine and for only one day.

Flowers small, blue or purple, with 6 equal obovate divisions: stigmas simple: stems or scapes flat or 2-winged, from fibrous roots; leaves narrow and grass-like,   (*Sisyrinchium*) BLUE-EYED-GRASS.

Flowers very large, orange and spotted with crimson and purple; the 3 inner divisions much smaller and narrowed in the middle: stigmas each 2-cleft: scape terete, from a coated bulb; leaves plaited,   (*Tigridia*) *TIGER-FLOWER.

Filaments separate: stigmas flattened, or petal-like.

Perianth 6-parted down to the ovary, regular and wheel-shaped, the divisions obovate-oblong, all alike, yellow, with darker spots: seeds remaining after the valves of the pod fall, berry-like and black, the whole looking like a blackberry (whence the common name). Stems leafy below, from a rootstock: leaves sword-shaped,   (*Pardánthus*) *BLACKBERRY-LILY.

Perianth irregularly 6-cleft; 3 of the lobes arched and making an upper lip, the 3
      lower more spreading, yellow, orange, or reddish. Stem rising from a
      corm, and bearing many flowers in a one-sided spike, (*Gladiolus*) *CORN-FLAG.

Perianth 6-cleft; the divisions of two kinds, the 3 outer recurved or spreading, the 3
      inner alternate with the others, smaller, erect, and differently shaped:
      stigmas 3, petal-like, one before each erect stamen. Generally with
      thick creeping rootstocks,   (*Iris*) IRIS.

Perianth with a slender tube, rising (with the linear flat leaves) from a corm or solid
      bulb (Fig. 76); the summit divided into 6 roundish, equal, erect, or
      barely spreading divisions: stigmas 3, thick and wedge-shaped, some-
      what fringe-toothed. Fl. in early spring,   (*Crocus*) *CROCUS.

Iris or Flower-de-Luce. *Iris.*

* Common cultivated species in gardens: outer divisions of the perianth with a bearded crest.

1. COMMON IRIS. Flowers several on a stem, 1° to 3° high, and much longer than the sword-shaped
    leaves, light blue or purple.   *I. sambucina.*

2. DWARF GARDEN IRIS. Flowers close to the ground, hardly exceeding the sword-shaped leaves,
    violet-purple, the divisions obovate, the 3 outer recurved. Fl. in early spring.   *I. pumila.*

* * Wild species.

3. CRESTED DWARF IRIS. Low and almost stemless, from rootstocks spreading on the ground; leaves
    short; flower pale blue, the tube thread-shaped (2' long) and longer than the spatulate divisions, the
    three outer divisions with a beardless crest. Fl. spring. S. and W., and in some gardens. *I. cristata.*

4. LARGER I. or BLUE-FLAG. Stem stout, 1° to 3° high, bearing several crestless and beardless purple-
    blue and variegated flowers, their inner divisions much smaller than the outer; leaves sword-
    shaped, ¾' wide. Wet places; flowering in late spring.   *I. versicolor.*

5. SLENDER I. or BLUE-FLAG. Stem slender; leaves narrowly linear (¼' wide), and flower smaller
    than in No. 4: otherwise much like it. Wet places, E.   *I. Virginica.*

102. ORCHIS FAMILY. Order ORCHIDACEÆ.

Plants with irregular and often singular-shaped flowers, the perianth standing as it were
on the ovary, as in the two preceding orders; but remarkable for having the stamens, only
one or two, united with the style or stigma. This may best be seen in the LADY'S SLIPPER,
of which we have three or four common species: the slipper is one of the petals, in the form
of a sac. The flowers of various sorts of ORCHIS are striking and peculiar; but the family
is too difficult for the young beginner, and therefore the kinds are not described here.
Fig. 69 represents two air-plants of this family, belonging to tropical countries.

III. Glumaceous Division.

103. RUSH FAMILY. Order JUNCACEÆ.

The true Rushes are known by having flowers with a regular perianth, which, although
glumaceous, i. e. like the chaffy scales or husks of Grasses, is of 6 regular parts, like a calyx,
enclosing 6 (or sometimes 3) stamens, and a triangular ovary. This bears a style tipped
with 3 stigmas, and in fruit becomes a 3-seeded or many-seeded pod. There are two

common genera, each with several species: the parts are too small and difficult for the young student.

Pod 1-celled and 3-seeded. Leaves flat and hairy,   (*Luzula*) WOOD-RUSH.
Pod 3-celled, many-seeded. Leaves generally thread-shaped, or none at all,   (*Juncus*) RUSH.

104. SEDGE FAMILY. Order CYPERACEÆ.

A large family of Rush-like or Grass-like plants, including the SEDGES, CLUBRUSHES, BULRUSHES, and the like, which have no perianth, but the flowers, collected in heads or spikes, are each in the axil of a single glume in the form of a chaff or scale. These plants are much too difficult for the young beginner.

105. GRASS FAMILY. Order GRAMINEÆ.

The true Grasses make a large and most important family of plants, with straw stems (called *culms*, 91) ; leaves with open sheaths; and flowers with 2-ranked glumes or chaffy scales, a pair to each flower, and another pair to each spikelet. It includes not only the very numerous kinds of true Grasses, but also of Corn, i. e. the Cereal grains, of which WHEAT, BARLEY, RYE, OATS, RICE, and MAIZE or INDIAN-CORN are the principal; also SUGAR-CANE, BROOM-CORN or GUINEA-CORN, and MILLET.

SERIES II.

FLOWERLESS OR CRYPTOGAMOUS PLANTS.

Plants destitute of flowers, and propagated by spores instead of seeds. See Part I., Paragr. 165, 308, 312 – 314.

CLASS III. — ACROGENS.

This class includes the FERNS, the HORSETAILS, and the CLUB-MOSSES.

CLASS IV. — ANOPHYTES.

This class includes the MOSSES and the LIVERWORTS.

CLASS V. — THALLOPHYTES.

Includes the LICHENS, the ALGÆ or SEAWEEDS, and the FUNGI or MUSHROOMS.

INDEX TO PART I.

AND

DICTIONARY OF THE BOTANICAL TERMS

USED IN THIS BOOK.

⁎ The numbers refer to the page where the term is explained or illustrated.

Bipinnatifid : twice pinnatifid.
Biternate : twice divided into threes.
Bladdery : thin and inflated.
Blade of a leaf, 43 ; of a petal, 64.
Border of a corolla, &c., 72.
Bracts and Bractlets, 59.
Branches, 24.
Breathing-pores of leaves, 264, 265.
Bristles : stiff and strong hairs.
Bristly : beset with bristles.
Budding, 56.
Buds, 24, 38.
Bulblets, 41, 57.
Bulbous : like a bulb in shape.
Bulbs, 31, 40, 57.

Caducous : dropping off very early, as the calyx of Poppies and Bloodroot.
Calyx, 7, 63.
Campanulate : bell-shaped, 72.
Capillary : slender and as fine as hair.
Capitate : headed ; bearing a round, head-like top ; or collected in a head, as the flowers of Button-bush, 61.
Capsule : a pod, 80.
Cartilagineous or Cartilaginous : like cartilage.
Caryopsis : a grain or seed-like fruit, 79.
Catkin : a scale-like spike, as of Birch, &c., 61.
Caulescent : having a stem which rises out of the ground.
Cells, in vegetable anatomy, 89.
Cells of the ovary or fruit, 8, 74.
Cellular Tissue, 41.
Cereal : relating to corn or corn-plants, held by the ancients to be the gift of Ceres.
Chaff : thin bracts, in the form of scales or husks.
Ciliate : fringed with hairs along the margin, like the eyelashes fringing the eyelids.
Circulation in plants, 86, 88.
Class, 94.
Classification, 93.
Claw, of a petal, &c., 64.
Cleft : cut about half-way down, 49, 50.
Climbing, 37.
Club-shaped : thickened gradually upwards.
Clustered : collected in a bunch.
Clustered Roots, 36.
Coated Bulbs, 40.
Coherent, calyx or ovary, 75.
Column : the united filaments of monadelphous stamens, as of the Mallow (Fig. 317), or

the stamens and style united, as in the Orchis Family.
Complete Flower, 67.
Compound Corymb, Cyme, &c., 63
" Leaves, 44, 51.
" Ovary, 73.
" Pistil, 73.
Compressed : flattened on two sides.
Cone, as of the Pine, 82.
Confluent : when two parts or bodies are blended together.
Conical Root, 36.
Connate : grown together from the first.
Connective, of the anther, 66.
Convolute, leaf, &c. : rolled up.
Convolute, in the flower-bud, 183, 187.
Cordate : heart-shaped, 48.
Coriaceous : of a leathery texture.
Corm, or Solid Bulb, 40, 57.
Corolla, 7, 63.
Corymb, 60.
Corymbose, or Corymbed : in corymbs, or like a corymb.
Cotyledons : seed-leaves, 9, 84.
Creeping, 57.
Crenate : the margin scalloped, 49.
Cruciform : cross-shaped, as the corolla of the Cruciferous Family, 124.
Crude Sap, 86.
Crustaceous : of a hard and brittle texture.
Cryptogamous, Cryptogamous Plants, 58, 97.
Culm : a straw-stem, 37.
Cuneate : wedge-shaped, 47.
Cupule ; the acorn-cup, and the like, 79.
Cuspidate : tipped with a sharp rigid point, 49.
Cut : said of leaves, &c., which appear as if cut or slit from the margin inwards, 49, 50.
Cuttings, 56.
Cyme, 62.
Cymose : in cymes, or like a cyme.

Deciduous : falling off, as petals generally do after blossoming, or leaves in autumn.
Declined : turned to one side, or to the lower side, 37.
Decompound : several times compound, 52.
Decumbent : reclined on the ground, 37.
Decurrent : said of leaves continued downwards on the stem, like a wing, as in Thistles.
Definite : uniform and rather few in number.
Dehiscence : the regular opening of pods.
Dehiscent Fruits, 79.

Dentate: toothed; the teeth pointing outwards but not forwards, 49.

Denticulate: toothed with minute teeth.

Depressed: flattened from above.

Diadelphous Stamens: united by their filaments in two sets, 73.

Dicotylédonous, Dicotyledonous Plants, 22, 97.

Diffuse: loosely and widely spreading.

Digestion in plants, 87.

Digitate, 51.

Diœcious Flowers, 68.

Dissected: cut into fine divisions.

Distinct: of separate pieces, unconnected with each other, 71, 73.

Divided: cut through or nearly so, 50.

Divisions, 49.

Double Flowers (so called), 69.

Downy: clothed with soft and short hairs.

Drupe: a stone-fruit, 78.

Drupaceous: like a drupe.

Dry Fruits, 77, 78.

Eared: bearing ear-like projections, or auricles, at the base, on one or both sides, 48.

Elaborated Sap, 87.

Elliptical: regularly oval or oblong.

Emarginate: notched at the end, 49.

Embryo: the germ of a seed, 6, 9, 83.

Endogenous Stem, Endogenous Plants, 41, 97.

Ensiform: sword-shaped, as the leaves of Iris (Fig. 64).

Entire: the margin even, not toothed or cut, 49.

Epidermis: the skin of a plant, 44.

Epiphytes: air-plants, 35.

Equitant (riding astride), 53.

Erect, 37.

Essential Organs of the Flower, 7.

Evergreen: holding the leaves green over winter.

Exogenous Stem, Exogenous Plants, 41–43, 97.

Exserted: protruded, or projecting, as the stamens in Fig. 45

Family, 94.

Farinaceous: mealy or like meal.

Fascicle: a bundle or close cluster, 63.

Fascicled Roots, 36.

Feather-veined, 46.

Fertile Flower, 68.

Fibrous Roots, 27, 36.

Fiddle-shaped: obovate but contracted on each side near the middle.

Filament (of a stamen), 7, 64.

Filiform: thread-shaped.

Fleshy Fruits, 77. — Plants, 31. — Roots, 35.

Floral: relating to the flower.

Floral Envelopes, 7.

Flower, 5, 7, 58.

Flower-bud: an unopened flower.

Flower-clusters, 59.

Flowering Plants, 58, 97.

Flowerless Plants, 58, 97.

Flower-stalks, 38, 60.

Follicle: a simple pod opening down one side (Fig. 210), 80.

Footstalk of a leaf, 43.

Free: not united with any other part, as when the calyx is not united with the ovary, nor the petals with the calyx, &c., 75

Fringed: the margin beset with bristles, &c., or finely cut into slender appendages.

Fruit, 5, 9, 77.

Fugacious: falling or withering very early.

Funnel-shaped, or Funnel-form, 72.

Generic name: the name of the genus.

Genus: plural Genera, 94.

Germ, 6, 9.

Germinate: to grow from the seed, 11.

Germination, 11.

Gibbous: projecting or bulging on one side.

Glands: a name given to very different things; to little fleshy bodies in some flowers (p. 128); to places in the leaves of the St. John's-wort, the Orange, &c., appearing like dots, which contain a volatile oil; and to the larger oil-cells in the rind of the Orange and Lemon. Also hairs or any projections on the surface of leaves or stalks which contain or exude any aromatic, glutinous, or watery matter, are called glands; as on the leaves and footstalks of the Sweet-Brier and of the Flowering Raspberry, p. 149.

Glandular: bearing glands, or gland-like.

Glandular hairs: hairs tipped with a gland or head.

Glaucous: whitish or whitened with a *bloom*, or fine powdery matter that rubs off, as that on a Cabbage-leaf.

Globose: shaped like a ball or sphere.

Globular: nearly globose.

Glomerate: collected into close or a head-like cluster.

Glumaceous: glume-like; resembling or bearing glumes.

Glumes : the chaffy bracts or scales which make the coverings of the flowers of Grasses, Sedges, &c.
Gourd-Fruit, 77.
Grafting, 56.
Grain, 78, 79.
Granular : composed of small particles or grains.
Growth, 89.
Gymnospermous (naked-seeded), Gymnospermous Plants, 76, 97.
Gynandrous : stamens borne on the pistil or style, as in the Orchis Family.

Hairy : bearing or covered with hairs, especially rather long ones.
Halberd-shaped, 48.
Hastate : same as halberd-shaped, 48.
Head, 61.
Heart-shaped, 48.
Heart-wood, 43.
Helmet : a name given to the upper sepal of Aconite (Fig. 254), &c.
Herbaceous, 37.
Herbarium : the botanist's collection of dried plants.
Herbs, 26.
Hilum : the scar of the seed, or point by which it is attached, 83.
Hirsute : hairy with stiff or beard-like hairs.
Hispid : bearing still stiffer and stouter hairs or bristles.
Hoary : grayish-white, or covered with a fine and close whitish down.
Hooded : shaped like a hood or cowl; concave or arched.
Horny : having about the texture of horn.
Hybrid : a cross between two species.

Imbricate or Imbricated : the parts overlapping; some of them outside and others inside in the bud.
Imperfect Flowers, 68.
Incised : irregularly and rather deeply cut, 49.
Included : enclosed; not sticking out.
Incomplete Flowers, 67.
Incurved : curving inwards.
Indefinite : too numerous to be readily counted, and not uniform in number.
Indehiscent : not splitting open, 78.
Indigenous : native to the country.
Inferior : growing beneath some other organ; as the calyx beneath the ovary, 75.

Inflated : bladder-like, as if blown up.
Inflexed : bent inwards.
Inflorescence, 58.
Inoculating, 56.
Inserted : borne on, or attached to, 71, 75.
Insertion : the place or the mode of the attachment of any organ to that which bears it.
Interruptedly pinnate, 52.
Inversely heart-shaped, 49.
    "   lance-shaped, 47.
    "   ovate, 47.
Involucel, 62.
Involucre, 62.
Involute : with the end or edges rolled inwards.
Irregular Flowers, or Corolla, &c., 71, 72.

Jagged, 49.
Jointed : separating by a joint, or dividing across into two or more pieces.

Keel : a projecting ridge on the under surface of a leaf, as of Day-Lily, &c. The two lower petals of a papilionaceous corolla united are also termed the Keel, or Keel Petals, 141.
Keeled : furnished with a keel or projecting ridge on the lower side.
Kernel of a seed, 83.
Key, or Key-Fruit, 78, 79.
Kidney-shaped, 48.

Labiate : two-lipped, 72.
Laciniate : slashed; cut into narrow and irregular lobes.
Lance-linear, 47.
Lance-oblong, 47.
Lanceolate or Lance-shaped, 46.
Lateral : belonging to, or borne on, the side.
Leaflets : the pieces of a compound leaf, 51.
Leaf-buds : buds which develop leaves.
Leaf-scars, 26.
Leaves, 6, 43.
Legume : a pea-pod, 80.
Limb of a corolla, &c., 72.
Lips, 72.
Linear, 46.   Linear-lanceolate, 47.
Lobed : having lobes, 49, 50.
Lobes : any strong divisions of a leaf, &c., 49.
Lower side of a flower : that which looks away from the stem, and towards the bract.
Lyre-shaped, a pinnatifid leaf with the end lobe largest and rounded, as in Radish (Fig. 57), 28.

Membranaceous: of the texture of membrane or thin skin.
Midrib: the middle rib of a leaf, 44.
Mineral Kingdom, 2.
Monadelphous, 73.
Monocotylédonous, Monocotyledonous Plants, 21, 22, 97.
Monœcious Flowers, 68.
Monopetalous: the corolla of one piece, 72.
Monosepalous: the calyx of one piece, 72.
Morphology, 34.
Mucronate, 49.
Mulberry, 82.
Multiple Fruits, 82.

Naked Flowers, 68.
Naked-seeded, 76.
Names of Plants, 94.
Napiform: turnip-shaped (Fig. 70), 36.
Natural History, 2.
Natural System, 96.
Naturalized: introduced from a foreign country, but run wild.
Nectariferous: honey-bearing.
Needle-shaped, 53.
Nerves, Nerved, 44, 45.
Netted-veined, 45.
Neutral Flowers, 69.
Notched, 49.
Nut, 78, 79.
Nutlet: a little nut or stone.

Obcordate: inversely heart-shaped, 49.
Oblanceolate, 47.
Oblique (leaves, &c.): unequal-sided.
Oblong, 46.
Oblong-lanceolate, 47.
Obovate: ovate inverted, 47.
Obtuse: blunt, 48.
Odd-pinnate, 52.
Offset, 39, 57.
Open Pistils, 76.
Opposite (leaves or branches), 25, 54.
Orbicular: circular in outline, 94.
Order, 94.
Organs, 5; of Reproduction, 5, 58.
    " of Vegetation, 5.
Oval, 47.
Ovary, 8, 65.
Ovate, 47.
Ovate-lanceolate, 47.
Ovules: rudimentary seeds, 8, 65.

Palmate, 51.
Palmately cleft, lobed, &c., 50, 51.
    "   veined, 46.
Panicle, 62.
Papilionaceous Flower or Corolla, 141.
Pappus: thistle-down, and the like; the limb of the calyx in the Sunflower Family, 165.
Parallel-veined, 45.
Parietal Placenta, 74.
Parted: cleft almost through, 50.
Parasitic Plants, 35.
Pedate: like a bird's foot; palmately divided, with the side divisions two-parted.
Pedicel: the footstalk of each separate flower of a cluster, 60.
Pedicelled: raised on a pedicel.
Peduncle: a flower-stalk.
Peduncled: having a peduncle.
Peltate: shield-shaped, 48.
Pepo: a gourd-fruit, 77.
Perennial: living year after year.
Perennials, 29.
Perfect Flower, 67.
Perfoliate: where the stem apparently passes through the leaf, as in Bellwort, No. 1 and 2, p. 211.
Perianth: the blossom-leaves, 64.
Pericarp: seed-vessel, 77.
Persistent: not falling off; remaining after flowering.
Petal: a leaf of the corolla, 9, 64.
Petiole: the footstalk of a leaf, 43.
Petioled: having a petiole or footstalk.
Phænogamous (also called Phanerogamous) Plants, 58, 97.
Pine-cone, 82.
Pinnate, 51.
Pinnately cleft, lobed, parted, &c., 50, 51.
    "   veined, 46.
Pinnatifid: same as pinnately cleft.
Pistil, 8, 65.
Pistillate Flowers, 68.
Pitcher-shaped leaves, 121.
Pith of a stem, 42.
Placenta, 66, 74.
Plumose: plume-like; feathered.
Plumule, 13, 84.
Pod, 79.
Pointed, 48.
Pollen, 7, 64.
Polyadelphous, 73.
Polycotylédonous, 22.

Spindle-shaped, 36.
Spiny or Spinose : bearing spines.
Spores, 58.
Spur : a slender hollow projection, as that of the upper sepal of Larkspur (Fig. 251), the lower petal of a violet (Fig. 73), &c.
Stamens, 7, 64.
Staminate Flowers, 68.
Standard of a papilionaceous corolla, 141.
Stellate : star-shaped.
Stem, 5, 23, 27.
Stemless : without a stem, or without one rising out of the ground.
Stemlet, 9.
Sterile Flowers, 68.
Stigma, 8, 65.
Stipel : the stipule of a leaflet.
Stipules, 43, 54.
Stock, 56.
Stolon, 39, 57.
Stoloniferous : bearing stolons.
Stone-Fruit, 77, 78.
Strap-shaped corolla, 165.
Strawberry, 81.
Striate : marked lengthwise with fine lines
Strobilaceous : resembling or bearing a
Strobile : a fruit like a Pine-cone, 82.
Style, 8, 65.
Subclass, 97.
Subfamily or Suborder ; a marked division of an order, such as might be considered important enough to form a separate order. See pp. 139, 146.
Subgenus : a marked division of a genus, such as might perhaps be taken as a separate genus.
Subulate : awl-shaped.
Succulent : juicy.
Sucker, 39, 57.
Suspended : hanging from the top.
Sword-shaped : erect and sharp-edged lance-linear leaves, like those of Iris (Fig. 64).
Superior : above some other part it is compound with, as "ovary superior," 75 ; on the upper side.
Symmetrical Flower, &c., 69.
Syngenesious, 73, 164.

Taper-pointed, 48.
Tap-root, 36.
Tendrils, 38.
Terete : long and round, like ordinary stems ;

same as cylindrical, but it may taper, as stems generally do.
Terminal : belonging to or borne on the summit.
Terminal Bud, 24.
Terminal Flowers, 52.
Ternate : in threes, or divided into three.
Ternately compound, &c., 52.
Thorns, 37.
Thread-shaped, 53.
Throat of a corolla or calyx : the summit of the tube inside.
Thyrse : a close compound panicle, like that of the Horsechestnut, 62.
Three-valved, &c., 80.
Thrice compound, thrice pinnate, &c , 52.
Tomentose : woolly, with a coat of soft entangled hairs or down.
Toothed : the margin cut into short and sharp projections or teeth.
Top-shaped : conical inverted, or with the point downwards.
Trailing, 37.
Trees, 27.
Triadelphous, 73.
Trifid : same as three-cleft.
Triple-ribbed : when a stout rib rises from each side of a midrib above the base.
Trumpet-shaped, 72.
Truncate : as if cut off at the end, 48.
Trunk, 37.
Tubers, 29, 40, 57.
Tuberous or Tuber-like Roots, &c., 36.
Tube of a corolla, &c., 72.
Tubular : tube-shaped, or with a tube, 72.
Tumid : swollen or thickened.
Turgid : nearly same as Tumid.
Turnip-shaped, 36.
Twice compound, 52.
" pinnate, &c., 52.
Twin : in pairs.
Twining : climbing by coiling, 37.
Two-lipped, 72.
Two-valved, &c., 80.

Umbel, 61.
Umbellet, 62.
Unarmed : not spiny or prickly.
Undershrub : a very low, shrubby plant.
Undulate : wavy.
Unsymmetrical Flowers, 70.
Upper : in a flower, the upper side is that next the main stem and away from the bract.

INDEX

TO THE

NAMES OF PLANTS IN THE POPULAR FLORA.

Jessamine, 189 · Linaceæ, 134 · Matthiola, 125
Jessamine Family, 189 · Linaria, 175, 177 · May-Apple, 120
Jewel-weed, 136 · Linden, 133 · May-flower, 169
Jointed-Charlock, 125 · Linden Family, 133 · Maypop, 155
Jonquil, 213 · Linnæa, 161 · Mayweed, 166
Juglandaceæ, 197 · Linum, 134 · Maywreath, 148
Juglans, 197 · Liriodendron, 117 · Meadow-Rue, 113, 114
Juncus, 216 · Lithospermum, 182 · Meadow-sweet, 147, 148
June-berry, 147 · Liverleaf, 113 · Medeola, 207
Juniper, 201, 202 · Liverworts, 216 · Medicago, 142, 144
Juniperus, 201, 202 · Lobelia, 167 · Medick, 142, 144
Kalmia, 169, 170 · Lobeliaceæ, 167 · Melanthaceæ, 209
Kentucky Coffee-tree, 148 · Lobelia Family, 167 · Melanthium, 209
Ketmia, 133 · Loblolly-Bay, 134 · Melilot, 142, 144
Knotgrass, 193 · Locust-tree, 142, 143 · Melilotus, 142, 144
Knotweed, 193 · Lonicera, 161, 162 · Melissa, 179
Koniga, 125 · Loosestrife, 152, 173 · Melon, 154
Labiatæ, 178 · Lophanthus, 179 · Menispermaceæ, 119
Labrador-Tea, 169 · Lopseed, 177 · Menispermum, 119
Laburnum, 142 · Lousewort, 176 · Mentha, 179, 180
Ladies' Eardrop, 153 · Lovage, 159 · Menyanthes, 187
Lady's Slipper, 215 · Lucerne, 144 · Mertensia, 181
Lagenaria, 154 · Lunaria, 125 · Mezereum, 195
Lamium, 180 · Lupine, 142 · Mezereum Family, 195
Lamb-Lettuce, 164 · Lupinus, 142 · Mignonette, 126
Lambkill, 170 · Luzula, 216 · Mignonette Family, 125
Laportea, 196 · Lychnis, 130 · Milk-Pea, 142
Larix, 201, 202 · Lycium, 186 · Milkweed, 188
Larkspur, 113, 115 · Lycopersicum, 185 · Milkweed Family, 188
Lauraceæ, 194 · Lycopsis, 181 · Millet, 216
Laurel, 169, 170, 171 · Lycopus, 179 · Mimosa, 143
Laurel Family, 194 · Lungwort, 181 · Mimosa Family, 143
Laurel-Magnolia, 117 · Lysimachia, 173 · Mimulus, 176
Lavandula, 178 · Lythraceæ, 152 · Mint, 179, 180
Lavatera, 131 · Lythrum, 152 · Mint Family, 178
Lavender, 178 · Lythrum Family, 152 · Mirabilis, 191
Leadwort Family, 173 · Maclura, 196 · Mirabilis Family, 191
Leatherwood, 195 · Madder, 164 · Mitchella, 164
Lechea, 127 · Madder Family, 163 · Mitella, 157
Ledum, 169 · Magnolia, 117 · Mitrewort, 157
Leek, 211 · Magnolia Family, 117 · Mockernut, 197
Leguminosæ, 141 · Mahonia, 119 · Mock-Orange, 157, 158
Lemon, 134 · Maize, 216 · Molucca-Balm, 180
Leonurus, 180 · Mallow, 131, 132 · Molucella, 180
Lepidium, 125 · Mallow Family, 131 · Mollugo, 130
Lespedeza, 142 · Malus, 147, 151 · Momordica, 154
Lettuce, 166 · Malva, 131, 132 · Monarda, 179, 180
Levisticum, 159 · Malvaceæ, 131 · Monkey-flower, 176
Lichens, 216 · Mandrake, 119 · Monkshood, 116
Ligustrum, 189 · Maple, 140 · Monocotyledons, or Mono-
Lilac, 189 · Maple Family, 140 · cotyledonous Plants, 97, 203
Lilium, 210, 212 · Marrubium, 180 · Monopetalous Division, 161
Lily, 210, 211 · Marsh-Mallow, 131 · Monotropa, 169
Lily Family, 209 · Marsh-Marigold, 113 · Moonseed Family, 119
Lily-of-the-Valley, 210 · Marsh-Rosemary, 173 · Morning-Glory, 184, 185
Lime-tree, 133 · Martynia, 174 · Morus, 196
Limnanthemum, 187 · Matrimony-Vine, 186 · Mosses, 216

THE END.